C000096789

Schering - Plough (Avondale) Co.
Rathdrum, Co. Wicklow Ireland.

RECEIVED

1 6 OCT 1991

Explosives in the Service of Man
The Nobel Heritage

Explosives in the Service of Man
The Nobel Heritage

Edited by

John E. Dolan
The Royal Society of Chemistry
Dolan Associates Consultancy, Ardrossan, UK

Stanley S. Langer
The Royal Society of Chemistry, London, UK

THE ROYAL
SOCIETY OF
CHEMISTRY
Information
Services

The proceedings of the symposium on Explosives in the Service of Man: The Nobel Heritage held on 9–11 December 1996 in Glasgow.

The cover photograph shows a bronze bust of Alfred Nobel held at the Ardeer site of Nobel's Explosives.

Special Publication No. 203

ISBN 0-85404-732-8

A catalogue record for this book in available from the British Library

© The Royal Society of Chemistry 1997

All rights reserved.

Apart from any fair dealing for the purpose of research or private study, or criticism or review as permitted under the terms of the UK Copyright, Designs and Patents Act, 1988, this publication may not be reproduced, stored or transmitted, in any form or by any means, without the prior permission in writing of The Royal Society of Chemistry, or in the case of reprographic reproduction only in accordance with the terms of the licences issued by the Copyright Licensing Agency in the UK, or in accordance with the terms of the licences issued by the appropriate Reproduction Rights Organization outside the UK. Enquiries concerning reproduction outside the terms stated here should be sent to The Royal Society of Chemistry at the address printed on this page.

Published by The Royal Society of Chemistry,
Thomas Graham House, Science Park, Milton Road,
Cambridge CB4 4WF, UK

Printed by Bookcraft (Bath) Ltd

Preface

This book is a record of the Proceedings of a Conference held in Tribute to Alfred Nobel on the 10th and 11th of December 1996.

The Centenary of Alfred Nobel's death on the 10th December 1896 was commemorated by a Symposium covering the science, technology and use of high explosives and a banquet in tribute to Alfred Nobel held in Glasgow close to the site where Alfred Nobel set up his British factory at Ardeer in Scotland.

The nineteen papers presented at the Symposium cover a wide spectrum of topics ranging from the historical background to both the past and present evolution of the science and a forward looking group of papers relating to the modern use of explosives in mining, quarrying, civil engineering as well as the more esoteric applications in the space programmes and forensic science.

John E. Dolan
Stanley S. Langer

Schering - Plough (Avondale) Co.
Rathdrum, Co. Wicklow Ireland.

Contents

General Introduction

Session 1. Historical

Session 2. Legislation and Safety

Session 3. Security and High Energy Explosives

Session 4. Explosives Design and Bulk Explosives

Acknowledgements

The Royal Society of Chemistry wish to acknowledge the assistance of the following people in organising the event.

The Organising Nobel 96 Committee

Mr. John E. Dolan	Chairman
Mr. P. McGoff	Institute Explosives Engineers (Rockfill)
Mr. W. Mather	Inst. Mining Engineers (ICI Explosives Europe)
Dr. J. Jeacocke	Explosives Industry Group (Exchem)
Mr. F. Murray	F.E.E.M.
Mr. V. Parker	ICI Explosives
Mr. S. S. Langer	RSC Industrial Affairs Division
Dr. H. Rendall	Paisley University
Mr. C. Campbell	Nobel Millennium Project -ASSET
Mr. M. Ball	
Mrs. E. Wellingham	Conference Secretariat

The Exhibition Team

Mr. A. Gray	ICI Explosives Europe
Mr. J. Herring	ICI Explosives Europe
Miss. A. Kelly	ICI Explosives Archives
Miss. S. Conway	ICI Explosives Archives
Mr. W. Ramsay	Kelvin Consultants.

The Publicity Team

Moira Donnelly	R.S.C. Public Relations
Nancy Guppy	RSC Press Officer
Mr. B. Henson	Press Liaison Glasgow

Exhibit Contributors

Mr. R. B. Hopler	Dyno Nobel U.S.A.
Mr. R. J. Crabbe	H. M. Senior Inspector of Quarries
Dr. M. Oglethorpe	RCAHMS Edinburgh

Thanks are due to the Session Chairmen:-

Session 1.	Dr. P. Bamfield
Session 2.	Dr. J. Jeacocke
Session 3.	Mr. F.M. Murray
Session 4.	Mr. G.E. Williamson

Session 5.   Mr. W. Mather
Session 6.   Mr. P. McGoff

Thanks are also due to all those who contributed papers, took part the discussions and who helped in any way in organising and stewarding the Exhibition.

General Introduction

Introduction

J. E. Dolan

37 SORBIE ROAD, ARDROSSAN KA22 8AQ, UK

Alfred Nobel the industrialist and the creator of the prizes which bear his name and the inventor of Dynamite was a most remarkable man. Born in Sweden in 1833 he was a man with an enviable ability to overcome adversity and to whom an obstacle was merely a target for achievement.

Much of his childhood was spent in a state of very poor health and in almost abject poverty. In spite of this he became an inventor of outstanding ability and was responsible for over 100 major inventions covering not only explosives, detonators and propellants but such things as celluloid film, artificial silk, leather cloth, synthetic rubber as well as working on biochemical studies on blood and new methods of telecommunications. In addition to all this he established and controlled fifteen major explosives manufacturing companies throughout Europe before he had reached the age of forty and eventually a world wide organisation covering 29 countries and some 90 operating plants.

The name Nobel is invariably associated with the invention of Dynamite. His most important invention was, however, the Detonator which he invented in 1863 and which has been hailed as the greatest discovery ever made in both the principle and practice of explosives. On it the whole of the modern practice of blasting has been built and still depends, a fact which is, in itself, a tribute to the genius of Alfred Nobel.

This two part invention - Detonator and Guhr Dynamite - gave to the world the era of the High Explosive and revolutionised mining and civil engineering.

The 10th of December 1996 marked the Centenary of the death of this remarkable man, the father of the modern High Explosives technology, and was commemorated in Tribute at a Banquet and Technical Conference held in Glasgow, Scotland some 25 miles from the site where he built his first Dynamite factory in the United Kingdom at Ardeer in 1871. The Banquet and the Technical Conference was held in the International Moat House Hotel at the Scottish Exhibition and Conference Centre in Glasgow on the 10th and 11th December 1996, coincidental with both the Centenary and the awarding of the Nobel Prizes in Sweden and was attended by some 140 delegates from places as far away as the USA, Australia, South Africa and India as well as from most European countries.

The programme of technical papers over the two days of the Conference was wide ranging, covering the Historical, Legislative as well as modern manufacture, modern blasting practice and, in today's context, the important developments in security control.

The Conference had a two-fold objective. Firstly to pay a proper and adequate tribute to Alfred Nobel on this unique occasion by reviewing the history of the development of Explosives, since the days of Blackpowder, and highlighting the vital importance of the contribution that the work of Alfred Nobel made to the modern era of High Explosives, to modern mining , civil engineering and construction industries. Secondly to carry that review into the latest technological developments including computer blasting simulation, the latest development in the new generation of electronic detonators, and civil and down hole oil well engineering as well as the more esoteric high technology uses in the Space exploration programme.

Central to the Tribute and Conference was a small but well presented Exhibition which displayed examples of Nobel's earliest manufacturing machinery. Backing the machine display was a screen pictorial panorama of the history of the development of the science of explosives from its earliest beginnings to the present day involvement of explosives in the space exploration programme. The Nobel Era was covered in two major panel displays, one covering the history and development at Ardeer,   the other covering Nobel in the United States.

A central feature of the Exhibition was a tableau showing the cartridging of Dynamite in the 1880s in which mannikins dressed in the authentic uniforms of the day were shown operating an original Dynamite hand cartridging machine.

The Exhibition was attended by school parties from local schools in the vicinity of the Nobel factory at Ardeer The Exhibition included two videos , the one showing *"The Life and Legacy of Alfred Nobel"* and the other an educational video on *"The Explosives Industry"*. Present Industry was also represented in the Exhibition with displays of manufacturers' current product range and technology.

The Banquet

The traditional Scots Banquet on the evening of the 10th , complete with piper, was preceded by a Civic Reception by the City of Glasgow and was attended by many dignitaries amongst whom were representatives of both Parliament and Government , the

Swedish   Embassy,   Industry and the Learned Societies. All paid tribute to the contribution that Nobel made to science, and to the encouragement of the sciences , medicine , literature and peace in the Noble prizes.

The formal toast to Alfred Nobel at the Banquet was made by Gudmar Johannes representing the Swedish Nobel Foundation. Mr. Johannes paid a most moving tribute to this great man who had done so much for the science of explosives, and for the encouragement of excellence in the sciences, the arts and peace with the Nobel Prizes administered by the Foundation set up after his death in accordance with his Last Will and Testament.

The Toast to Alfred Nobel by Gudmar Johannes

To most people, Alfred Bernhard Nobel was, and will remain, an enigma.

He   was a man of sorrows, well acquainted with grief. Standing on the site   at Heleneborg in Stockholm, where - on the 3rd of September 1864 - the disastrous explosion killed 5 people , among them, his youngest brother, Emil, one wonders   why did he not stop at this stage - and once and for all give up experimentation   with Nitroglycerine? What forced him to continue? Was it the instinct of a dedicated scientist, coupled   with a strong, personal belief in himself to overcome any difficulty or problem? Or was he simply clutching at straws, desperate to succeed and make it possible for him to compensate for all the sufferings the Nobel family had gone through, and to be able to repay his parents for all they had given him in such abundance - love, encouragement and support ?

One of the more remarkable achievements of Alfred Nobel's life is the fact that, a mere few weeks after that devastating accident, he managed to establish at Vinterviken outside Stockholm, the world's first company for the commercial production of Nitroglycerine, "The Nitroglycerine Company", later to be known as "Nitro Nobel", and laid down the foundation of an astonishingly successful career.

However, in spite of that success he remained a lonely, melancholy man, plagued by a bad stomach, a frail heart, bedevilled by opposition and set-backs, but most of all by missing a life companion.

Yet, contradictory as history and the human character can be, Alfred Nobel has been described as a man of great humour, a most entertaining conversationalist, socially gifted, a brilliant linguist and above all - as a man who was generous almost to a fault! There are countless reports on how he cared for - besides his closest and dearest relatives - his colleagues, his workers, his servants and also people whom he met casually and who aroused his interest and to whom he took a liking.

This is illustrated in the following, very personal story that has never been told before. It is a story which demands an appropriate occasion and this Centenary is that occasion; this is the place and this is the audience!

The story involves my own personal encounter with someone who had had the privilege of meeting and speaking with Alfred Nobel in person. The encounter took place in 1985 on a train between Geneva and Paris. A ninety-seven year old lady - born in 1861 - also travelling to Paris got into conversation with me. She had heard of my interest in Alfred Nobel and she told me the following story.

"The year was 1877. We were a party of pupils from several countries travelling from a girls' boarding school in Geneva where we were studying humanities. I was with a group of 16 year old Swedish girls travelling together in the one carriage. As part of our studies we were supposed to learn French and to the extent that we were obliged to speak only French, which study was the reason for this journey to Paris. On board the train, however, we were allowed to relax and speak our own languages for a couple of hours. We had lapsed into our native Swedish tongue and were engaged in conversation when a man passing through the carriage stopped when he heard Swedish being spoken and asked if he could join us. I was totally unaware at the time who he was but I vividly recall he was dressed in a black frock coat and holding a top hat.

"He showed great interest in our studies and asked in particular about our reading and literary preferences. Before he left us, he revealed his own predilection for poetry, by quoting some lines in English to us. To judge from his refined and cultured attitude and demeanour , we believed he was probably in a scholarly or artistic profession but was someone who, from wide ranging, interesting, talk had obviously spent long periods of time abroad.

"Upon our arrival in Paris, our school-mistresses took us - surprisingly - to an elegant restaurant, where the entire group of boarding school pupils and teachers were treated to a superb lunch at the stranger's expense. He, himself was not present but our teachers told us, that his name was - Alfred Nobel, and this was his way of thanking us for making his journey so pleasant.

"Later, at the age of 21, I married a nobleman of Scottish ancestry , Carl Gustav Spens, who was just at the start of a great career with the Swedish National Defence. He was descended from an old, distinguished Scottish family, whose lineage connected with the 'Thanes of Fife', and dated back to the 13th century and whose descendants had come to Sweden in the 17th century. Marriage to Carl-Gustaf Spens , among other things, brought me into contact with Alfred Nobel's work and businesses in Scotland, and I gradually became familiar with the personality of the stranger that I and my schoolmates met on that train journey to Paris all those years before.

"By one of those strange co-incidents in life, again through my husband's Scottish connections I, at the same time, also became interested in the fascinating and sensitive poetic world of Robert Burns. I had often wondered, what poem did Alfred Nobel actually read to us on the train that day? From his reading I remember one single word out of the content, the word 'briethers'.

"I eventually found the word in a poem by Robert Burns which I am convinced is the verse which Alfred Nobel read to us on that train journey so long ago. The subject, the style, the words, everything corresponded so much to my recollection of that emotional recitation on that journey. "

The old lady then handed me a booklet of verse in which she had marked the immortal words:-

> "Man to Man the world o'er
> Shall brothers be for a' that!"

The story of this meeting and conversation confirms my belief and opinion, that Alfred Nobel, poet and philosopher that he was and living here in the west of Scotland where he started the process of creating one of the most important cornerstones in his dynamite empire, the British Dynamite Company, must have known and loved the genius of Robert Burns, and here in Scotland is the fitting place and this the Centenary the fitting occasion to tell this story for the first time and to quote that verse in full.

I am extremely grateful to my sister-in-law, who arranged for me the meeting with her grandmother - the old woman - who passed away only a year afterwards.

> "Then let us pray that come what may
> As come it will for a' that
> That Sense and Worth, o'er a' the earth
> May bear the gree and a' that
> It's comin' yet for a' that
> That Man to Man, the world o'er
> Shall Brothers be for a' that!"

It is doubly appropriate in this, the Bicentenary year of Robert Burns and the Centenary of the death of Alfred Nobel to link these two men of international fame both of whom were, although so different in personalities, brothers both as poets and philosophers.

So, my Lords, Ladies and Gentlemen, I am proud to ask you to please be upstanding and join in a toast to the memory of Alfred Bernhard Nobel - Scientist, Businessman, Philosopher and Poet!

The reply on behalf of the Chemical Industry was made by Sir Ronald Hampel, the chairman of ICI. Sir Ronald in particular drew attention to lessons which management in modern industry could learn from this role model combining, as he did, his brilliance as a practical scientist, with an equal brilliance in international finance and business management.

The reply on behalf of the Chemical Industry by Sir Ronald Hampel

The chemical industry owes a debt of gratitude to Alfred Nobel not so much for the Nobel Prizes as for the invaluable example which his life challenges others to follow. Nobel is the perfect example of the kind of man that British companies should be trying to develop: an innovator, a true European and a global businessman.

Nobel was in fact a rare combination - inventor, innovator and businessman but above all an Innovator.

The essence of innovation is not so much the invention itself. Innovation is the commercial and profitable development of new ideas or inventions. Innovation is to do with markets. It is identifying a market before designing, sometimes inventing, the right basic product, supplying your market, then producing an endless stream of new developments to keep your customer wanting for more.

Nobel exploited that concept almost to perfection. During the course of his lifetime, he was responsible for over 350 inventions - a variety of electric detonators, ballistic powders, detonating cords, a great range of blasting agents - nearly all of which he exploited commercially and built into a world-wide business that kept growing rapidly until his death 100 years ago today. By the time he died in 1896, there were over 90 Nobel factories around the world.

The lesson is there, but how many UK companies and industries have gone under through failure to innovate? Britain has always been good at invention, and still is. But as innovators, compared with Japan, USA, Britain is still in the dark ages. How many inventions have been surrendered to other countries because we failed to take any of the important developmental steps? Classic examples are the jet engine, the body scanner and the silicone chip. Each of these were invented in UK, but developed and marketed by others.

Britain is still in the forefront of discovery and invention - only the US has won more Nobel prizes. Indeed this year, Britons have been awarded two prizes, and one is for chemistry - to Sir Harry Kroto of the University of Sussex for his work on Carbon 60 (Buckminsterfullerenes). Work on the exploitation of this discovery is gathering speed with exciting possibilities in the fields of pharmaceuticals, materials, high temperature super-conducters and optoelectronics, but will the commercial rewards come to Britain? Or will other more adventurous nations once again gain the benefit?

Chemical Industry in Britain would do well to take heed of what Kroto himself says:-

> "This discovery (C60) stands as yet another shining example of the capacity of non-strategic research programmes to make major advances. It thus stands as a timely warning of the serious limitations inherent in the applied research strategies which have always dominated the financial support of (British) science, and now appear to be taking over altogether".

Nobel taught us another lesson - the advantages of being European. Geography and sentimental nationalism never stood in the way of Nobel pursuing growth and excellence. By the age of nine, after living in both Sweden and Finland, he was at school in Russia at St Petersburg and spent two or three years in America before he was twenty one, followed by Berlin, Paris and London before he was much older, fund-raising for the business - in the days when travel really was travel!

Nobel spoke Swedish and Russian of course; he also spoke fluent English, French and German. How many British executives trouble to learn even one foreign language fluently - or companies to train them?

It is clear that European borders meant little to Nobel. The whole of Europe was home to him. It is indeed interesting to speculate on what his views might have been about the European Union, given that he lived in at least 6 European countries, and spoke 5 languages fluently, the probability is that he would have been amazed at the continuing insularity of Britain in Europe. He would have totally understood the need to be a leading player and influential negotiator at the table. He would also have recognised the need for Europe as a whole to be competitive against world markets.

The third lesson his life teaches the business man of today is the vital need to be global. It may seem obvious to us now, but 130 years ago, few people thought in terms of a global market. Nobel did. He thought further. He thought in terms of world peace and prosperity hence the Nobel prizes, not just for science, but for the humanities. They remain the oldest and most important international prizes for achievement.

Nobel was the first to create a multinational business at a time when most other companies were only just learning to operate nationally.

The first managers of ICI and their successors, steeped as they were in the Nobel tradition, unlike too many other British companies, followed the Nobel principle and drove ICI globally - and still do. It is that approach that has kept ICI alive when other historically great British industries have all but disappeared.

Today we face global competition in way even Nobel never conceived. There is an overriding need for innovation, and a strong science base, not just in Britain but in Europe as a whole. A fundamental research base is essential for a strong single market and a proper European domestic market will provide a home base from which to compete world-wide. Nobel would have wholly approved.

Nobel had a totally unjustified modest opinion of himself and of what he had achieved. In his own words his greatest merits were:-

> "To keep my nails clean and never be a burden to anybody.
> My greatest fault: to lack a family, good temper and good
> stomach. My greatest and only demand: not to be buried
> alive. Important events in my life: none!"

But even in that modesty there is a lesson to be learned because it can be and was in his case the driving motivating force to do better and improve on what has already been achieved and never accept "impossible" as an answer in spite of the difficulties and setbacks and vicissitudes of life.

The Chemical Industry owes Nobel a deep debt of gratitude not just as a scientist and the founder of the Nobel Prizes but as a silent tutor by example of business methodology. It is, tonight, a pleasant duty to have the opportunity to acknowledge that debt and to express the thanks of the Chemical Industry for the lessons he taught.

––––––––––––

The Vote of Thanks on behalf of the guests was made by Mr. Peter McGoff, the President of the Institute of Explosives Engineers who, in particular, praised and thanked the speakers for their most excellent Tributes and the Nobel 96 Committee for the organisation of the event.

––––––––––––

The Banquet was highlighted by the formal presentation of a bust of Alfred Nobel to the President of the Industrial Affairs Division of the Royal Society of Chemistry Dr. Peter Bamfield by Sir Ronald Hampel in commemoration of the occasion.

The presentation of the Bust of Alfred Nobel to The Royal Society of Chemistry by Sir Ronald Hampel Chairman of ICI.

No name is more familiar to generations of ICI people than Alfred Nobel.

Mond and the great Harry McGowan have both been remembered over the past few days because, coincidental with the hundredth anniversary of Nobel's death, only last week ICI celebrated the seventieth anniversary of its foundation. It is interesting that while there are busts of McGowan and Mond outside the head office in Millbank, nobody really notices them because they are high up on the building, Nobel is seen, however, every time you go in through the entrance to IC house, on Millbank or up to the Board Room at Ardeer - you can't   miss him. Nobel remains the scientific and business inspiration behind ICI.

To celebrate the hundredth anniversary of his death, and the great legacy that he left, ICI has the privilege of presenting one of the marble busts of Alfred Nobel, which has graced the halls of ICI for many decades, to the Royal Society of Chemistry in honour of this occasion and in recognition of their initiative in suggesting and being a principal in organising this conference in Tribute to the great man.

You may wonder why the bust here is bronze, and not marble. There is a simple and practical explanation and one which would have appealed to Nobel. The bronze bust requires only 4 men to shift it! This particular bronze bust actually belongs to Nobel's Explosives at Ardeer, just down the road, and they hope to get it back. The marble bust, on the other hand, weighs one and a half tons, and is down in London! It was simply a matter of practical logistics. Yes, we did think of painting it white for the occasion, but decided not to. So I hope you will not mind if this presentation is in a sense symbolic and trust that many of you will be able to see the real thing in its new home in the entrance foyer of Burlington House.

FIGURE 3.
SIR RONALD HAMPEL   ON RIGHT, DR. PETER BAMFIELD ON LEFT.

Epilogue

The event was deemed by all to be a success and a proper tribute to a great man of science and, perhaps, an example of how the importance of science in our everyday lives should be acknowledged by those in industry who owe a debt to that contribution.

This Conference was organised by the joint learned societies of:-

THE SOCIETY OF CHEMICAL INDUSTRY,
THE INSTITUTE OF EXPLOSIVES ENGINEERS,
THE INSTITUTE OF MINING ENGINEERS
THE EXPLOSIVES INDUSTRY GROUP (UK).

And further sponsored by:-

I.C.I. EXPLOSIVES EUROPE
EXCHEM EXPLOSIVES PLC
IRISH INDUSTRIAL EXPLOSIVES
FEDERATION OF EUROPEAN EXPLOSIVES MANUFACTURERS
ARDROSSAN SALTCOATS AND STEVENSTON ENTERPRISE TRUST
GLASGOW DEVELOPMENT AGENCY
THE EMBASSY OF SWEDEN

JOHN E. DOLAN
Chairman Nobel 96 Committee and Conference Chairman

Session 1. Historical

The Gunpowder Era

J. Jeacocke

EXCHEM EXPLOSIVES PLC, COMMONWEALTH HOUSE, NEW OXFORD STREET, LONDON, UK

The origins of GUNPOWDER or BLACKPOWDER are lost in the mists of time. It is reputed to have been known to the Chinese over 2000 years ago. In spite of this very long history, there is no recorded use of Blackpowder in practical mining or quarrying before the seventh century. Before then it had a varied and mystical career in which the religious, warlike and the entertaining were all intrinsically mixed up.

Until about the tenth century there was no clear idea about the correct proportions of nitrate, charcoal and sulphur to use, and the preparation was surrounded in mysticism. The association of charcoal and sulphur with the evil spirits and the satanic, together with the ancient rites of fire worship sustained the association with the mystical and led to the widespread use of Fireworks to dispel evil forces, a practice which is still very much a part of life in China today.

This connection between Blackpowder and the ecclesiastical community seems to persist throughout its history even to this day. The Franciscan monk Roger Bacon is reputed to have been the first to have made systematic research into the composition of Blackpowder followed shortly by the Abbe Schwartz and the connection between the two names will not be lost on linguists. Legend has it that the monks at Waltham Abbey were making Blackpowder in the 1500's and in the USA today there is a priest called " the master pastor blaster" while in the UK there is the Reverend Ron Lancaster. There is more than a suspicion that the association of " fire and brimstone" has a lot to do with it; there is no doubt that once one has got the smell of sulphur in one's nostrils one is "hooked".

The first reference to the use of Blackpowder for practical purposes is in the military field where Marcus Greacus in the year 700 A.D. describes its use in crude rockets and thunder flashes for demoralising the enemy. The military application dominated the development in Europe from the 13th century. By the 15th century an industry had grown around Blackpowder for military application but still no use had been made for mining purposes. This had to wait for a further two centuries. It was, in fact, first used for mining in Hungary in the 17th. century. The potential was immediately recognised and the use soon spread to Germany and Britain.

The development into the use for mining may now seem obvious but, at the time, it was a brilliant piece of inventive engineering which has, in the succeeding centuries revolutionised mining, quarrying and civil engineering practice and undoubtedly significantly increased the rate of scientific progress. The earliest attempts at using Gunpowder for blasting consisted of simply pouring the powder into natural fissures and breaks in the rock. The idea of preparing special shot-holes followed very quickly and were recorded in use by 1637.

These early shot-holes were made by iron-tipped borers driven in by jack hammers and closed with wooden wedges called "shooting-plugs" in an attempt to retain the gasses and develop the explosive's effect. However by 1685 clay stemming was being used for this purpose with much greater effect.

During the 18th and the first part of the 19th century firing of gunpowder shots strikes one as being a somewhat risky procedure. A charge of gunpowder was poured into the shot-hole and a long needle or stick inserted into the charge . The bottom of the stick was in the charge and the top protruded from the top of the shot-hole. After the stemming with clay had been carried out, the stick was withdrawn and the hole so left was carefully filled with loose, fine-grained gunpowder. A piece of touch paper (which was supposed to take about half a minute to burn but seldom did) was lit and laid on the gunpowder trail. The final, and from the shot-firers' point of view, the most important act of the performance was for the shot-firer to run as fast as possible.

This original crude method was vastly improved when William Bickford of Cornwall invented the SAFETY FUSE. The final refinement to the system of Fuse Firing was applied when, in 1804, Baron Chastel of Austria showed that a series of safety fuse primed charges could be ignited by electric sparks. At this point modern technology steps into the picture for the first time.

Turning now to the essential ingredients, charcoal, salt petre and sulphur. Much of the success in the use of Blackpowder depends on the charcoal - assuming that the sulphur and nitre are reasonably pure. Charcoal was originally burned by the classical method of building a wigwam of sticks, covering it with inverted turves and setting light to it. A pre-war interviewer on the radio asked a Forest of Dean charcoal burner how he knew what was going on inside. The reply, which shocked the establishment at that time, was "I spits on the outside and watches it sizzle". Sulphur also was very much of an unknown variable; sulphur-bearing rock was mined and the sulphur produced by distillation. The coming of the Frasch process brought sulphur of 99% plus purity enabling quality control to be increased. As for salt petre, the less said about it the better. There has been much controversy in the past over the ownership of the crystalline deposits found on the walls of manure pits under stable yards; once again the art was in the purification although the more easily obtainable sodium nitrate from guano was sometimes substituted.

The Strand Magazine of 1895 Vol. 9 pages 307-318 describes a visit to the Royal Gunpowder Factory at Waltham Abbey and gives a graphic description of the refining processes and the manufacture of charcoal.

Gunpowder mills were invariably located where there was a source of water power, the only available means of energy, it being realised very early on that incorporation in a pestle and mortar produced a much inferior powder to that from a stamping mill and even that from an edge runner mill.

The Strand Magazine referred to even shows the first Safety features, a "flash board" which, in the event of an explosion tipped backwards and discharged tanks of water onto the charge and those in the other mills. Before engaging the gears the operator " prudently draws down the flaps of his cloth helmet" (deerstalker!) and retires.

Such was the awe in which Black powder was held that, as far back as July 1776 the following instruction, which makes very interesting reading, was issued regarding precautionary safety procedures in the handling of all kinds of gunpowder.

"Whosoever is at Labour within or without the powder magazines should execute his commission in such a respectful and reverend silence as is seemly in such a place where (unless the almighty in his Grace keeps a protective hand over the Labour) the least lack of care may not alone cause the loss of life of all present, but may even in a moment transform this place as well as its surroundings into a heap of stone. Everybody is charged with the utmost caution and prudence in the handling of the powder by due observance and remembrance of what hath been deemed to that end. Furthermore, all, whether employed in the mixing of the powder or in the transport thereof, be it out of annoyance at Labour or still less out of lack in faith are most earnestly beseeched not to let emane from their mouths oaths or swearwords or other light or obscene language, whereby the Name of our Lord is dishonoured and taken in vain; as those who are themselves guilty thereof shall without tolerance or apology immediately leave their jobs and be delivered into the hands of the sentinel until the Labour is ended. Whereupon they are to be put under arrest and in accordance with the verdict be sentenced for the crime committed."

This instruction could well have emanated from the gunpowder mills at Waltham Abbey where records show that the first gunpowder mill was in existence in 1561. From 1702 to 1787 the "Mills" were owned by the Walton Family (of Sir Isaac Fame) who sold them to the government in that year for £10,000.

Notwithstanding that there were a number of gunpowder mills supplying the government and calling themselves " Royal", Waltham Abbey was the first truly " Royal" and thus the first nationalised industry.

In fact, by this time, the whole industry was in such chaos that, notwithstanding earlier legislation, the "Gunpowder Act", of 1860 was enacted to try to regulate the industry and reduce the number of accidents.

An interesting sidelight upon this concerns the establishment of Woolwich Arsenal. There has always been an argument as to which was the senior establishment.

When it finally emerged that Woolwich was set up with the proceeds from the sale of some " bad gunpowder", Woolwich countered by saying that everyone knew the source of the bad gunpowder - so much for solidarity.

Quite apart from the safety problems in manufacturing, transportation of the finished product brought problems of its own. The following are the recommendations (seemingly made solely for the port of London) which are given in full if only for their quaintness, particularly the Twelfth.

"The following is therefore the best mode of avoiding such accidents as have so frequently happened, and are likely to happen, from the gross inattention to the conveyance and storage of gunpowder.

First, - All wooden barrels filled with Gunpowder should have the word "Gunpowder" distinctly marked on them in black letters of at least an inch long.

Second, - All wooden barrels filled with Gunpowder, carried through the streets, should be wrapped up in leathern bags or fearno't bags, (but not in saltpetre bags, as allowed by the Act of Parliament, as they are a most useless covering,) which should be marked with the word "Gunpowder" in letters at least an inch long.

Third, -All Carriages in which Gunpowder is conveyed from the Mills should be hung on easy springs, and should be covered with wooden tilts and should be nailed, or put together with copper nails, in those parts where the Powder is likely to be spilled; they should be made quite close, and to open only by a door behind, hung on brass hinges, which should be stuffed with wool, so as to prevent the jolting of the carts from shaking the Powder out of the casks.

Fourth, - No carmen or watermen, whilst employed in conveying Gunpowder, should be allowed to wear iron nails in their shoes.

Fifth, - No Gunpowder should, under any pretence whatever be carried through the streets of London, for the purpose of being shipped off to the sloops or magazines or for exportation, but should invariably be sent by water from the nearest ferry to the Mills.

Sixth, - No Gunpowder exceeding 200lbs., the quantity allowed to retailers, should be shipped off from, or brought to, any stairs or landing place above Bow Creek, on any account.

Seventh, - Bow Creek will not only be a safe place, but will be a most convenient place for landing Powder for the town trade, as from thence the carts can convey it to the different inns and dealers, with nearly as much ease as they now do from Wapping, and without any risk.

Eighth, - Blackwall being no longer a proper boundary for powder sloops, all ships and vessels of every description should be compelled to deliver their store powder on their arrival, and to receive it outward bound below Bow Creek.

Ninth, - The boats which receive Gunpowder from the carts, to take it to the magazines or ships, or from the magazine or ships, to convey it to the Mills, should not only be decked, but marked with the word "Gunpowder" in large letters.

Tenth, - The powder sloops, notwithstanding the Act directs to the contrary, seldom are without fires in winter, which should not be permitted on any account, whilst partly or wholly loaded with Gunpowder; the penalty for so doing should be, at least, confiscation of the boat, or sloop, and their cargo.

Eleventh, - in order to prevent the frauds practised by watermen who take Gunpowder from ships on their arrival, and the risk that is likely to happen from such a class of persons having it in their possession, a certain number of watermen, say fifty, should be appointed (under the discretion of the Corporation of the City of London, who should take ample security) to put it on board, or receive it from the ship, or take it to or from the magazines, and none other should have charge of it.

Twelfth, - As the accidents that have happened by explosions of Gunpowder are innumerable, and as Wooden Barrels, in spite of every care, are an unsafe conveyance, it is strongly recommended that all retailers of Gunpowder, all persons working of quarries, and all ships of war and vessels of every description, should keep their Gunpowder in Barrels of Wood or Copper Cylinders; and that all Wood Barrels should be close joined and full bound round with Hoops to secure them; and that all copper Cylinders should also be bound round with a least three hoops; and that all Wooden Barrels and Copper cylinders (not so secured), containing Gunpowder, and their contents, together with the wagon, cart, or other conveyance used in transporting such Gunpowder, shall, without exception, be sizeable.

Nevertheless, in such carriages, and under the charge of such people, are thousands of barrels of Gunpowder moved from the Mills at Hounslow, through the different town into London, by Piccadilly, along the Strand, Fleet Street, Past St. Paul's, the Bank, the India House, the Tower, and last of all by the London Docks, to Union Stairs, Wapping, where the carmen, with iron nails in their shoes, get into the wagons and carts, in order to unload them, which, by the time they arrive, are often strewed with Gunpowder from one end of the floor to the other, jolted out of the barrels through the saltpetre bags. That neighbourhood has hitherto escaped an explosion, which certainly would be destructive to all around; there these thoughtless carmen, in the midst of drunken sailors of all nations, and the most wretched and abandoned women, drinking and smoking, unload their carriages subject to accident from fire, and the wickedness of ill-disposed people, and carry the Powder to the waterside through a passage, into which a door of a public-house opens, surrounded by the lowest orders of society and watermen smoking and priming their boats, there are boats appointed for the purpose receive it; and the saltpetre bags are, contrary to the Act of Parliament, taken off the barrels, and returned to the wagons. The Act directs also, that the boats should proceed immediately down the river, which is rarely completed with; and it is a fact, that these small boats are often long on their passage to the sloop magazine or ship, and are very frequently and improperly managed by boys, who lay them most imprudently alongside houses, ships, the entrance of the docks, & c. &c. On the 27th January last, the sloop *"Success"*, lying above Blackwall, was with two boys, very nearly blown to atoms, in consequence of her being left by the master, with a fire on board, (in direct opposition to the Act of Parliament), by which means two boys, (who had got possession of one pound cartridge of Powder) in making squibs, exploded it and were nearly killed, and nine barrels of Powder, each containing 100lbs had nearly exploded, the force whereof, from the improper situation in which the sloop was, might have done irreparable mischief.

Storage was the next problem and the positioning of magazines was the subject of much dispute. The early Safety Distances were as defined as follows.

> The store shall not be within the city of London or Westminster or within three miles of either of them, or within any borough or market town or one mile of the same, or within two miles of any palace or house residence of His Majesty, his heirs and successors, or within two miles of any gunpowder magazine belonging to the Crown, or within half a mile of any parish church."

Today, the commonest use of Blackpowder is in the manufacture of fireworks. It is interesting that display shells are fired from a " mortar". History suggests that the first notice of this came when powder mixed in a mortar with a pestle was left loosely covered with an iron plate. A stray spark from somewhere ignited the contents and the plate was projected some way, this is probably apocryphal but the term has persisted - even to the military " trench mortar"

In earlier times, cannons of the Gunpowder era were , of course, muzzle loaded with the main charge and a trickle of a finer grade loose powder poured down the touch hole from whence comes the term " spiking the gun" by driving a nail into the touch hole preventing the gun being fired.

As mentioned above rockets were first seriously referred to by the Romans. Rockets have developed in two different ways; the fireworks rocket and the war rocket The modern generation of war rockets was conceived by the second Baron Congreve, developed at Waltham Abbey and first used in the Battle of Copenhagen in 1801 and accounts for the line in the National Anthem of the USA which refers to the "rockets red glare". But it is on life saving that the rocket has made the greatest impact. The early "ship to shore" line throwing rockets were obviously gunpowder powered as they were exhibited at the Diamond Jubilee Exhibition of 1897 when they were awarded a Gold Medal.

The most famous illegal use of Gunpowder was, of course, that of Guy Fawkes; the story is too well known to be repeated here. It is interesting though that it was not until 1978 that the official records of Nov. 7,1605 were examined to show that 1800lbs of gunpowder was removed from the house and conveyed to the magazine at the Tower. Whether or not the transport regulations referred to earlier were obeyed is not clear, but the exercise cost 15s and 6d.

Many are the obvious uses to which gunpowder has been put, indeed, it was even used to power a pile driver but it is not recorded how many times it worked. Even Her Majesty's Customs and Excise used it to "prove spirit" for revenue purposes.

Finally reference must be made to the man who, in 1963, had the revered title of "second Inspector" who bequeathed to posterity a poem which elegantly describes Gunpowder in the process of detonation.

Initiators fire the chain
Acceleration boards the train
Fierce and fast reactions zip
Ingredients, self-sufficient, whip
The pace beyond "combustion"
Past the point of "Deflagration"
Atoms fly with mounting pressure
"Explosion" then becomes the measure
Ah! but for some that's not enough
For they are made of rougher stuff.
And still the pace goes up and up
Until it reaches ceiling, top
The pace by now extremely hot
No more acceleration can be got
Energy loosed in shocking wave
Atoms agitated so behave
With truly violent reputation
The label then is "Detonation".

The Nobel Era

J. E. Dolan

37 SORBIE ROAD, ARDROSSAN KA22 8AQ, UK

The Industrial Revolution of the 19th century saw the birth of the age of technology, the key to which was power and without which civilisation as we know it could not be sustained.

It is little wonder then that so much scientific and engineering research has been spent on exploring new fields of endeavour from which to generate this all important resource.

Within this millennium power has been obtained from the movement of water, from solar energy, from coal, from chemicals and other fuels, and now from atomic energy. In the array of power sources that are available today, that obtainable from chemical energy in the form of high explosives is and will continue to be, one of the most dramatic.

It is, however, a sobering and intriguing thought that this most sophisticated form of chemical energy is, fundamentally, the same as the oldest form of energy known to man - fire.

The art of the explosive technologist over the centuries has been to devise ways of increasing the speed of this reaction to the point where it takes place faster than the speed of sound in its own material, at which point the reaction escalates into a shock front. It is the shock wave associated with supersonic chemical reactions that is the target of the explosive technologist.

Prior to 1850, man's use of chemical energy in mining, was limited to those tasks which could be reliably performed by the deflagrating explosive, Blackpowder. Blackpowder being a reaction between two separate chemicals is an inter-molecular reaction and dependent for its explosive efficacy on the degree of confinement so that its use in mining and particularly civil engineering was strictly limited.

The final breakthrough to the High Explosive depended on finding a technique whereby the high speed regime could be sustained without the necessity for confinement. This step was accomplished by effectively carrying out the combustion reaction as an intramolecular reaction within the same chemical molecule.

This step was achieved with the discovery of the molecular explosive Nitroglycerine - Glycerine Trinitrate

$$CH_2\text{-}ONO_2$$
$$|$$
$$CH\text{-}ONO_2$$
$$|$$
$$CH_2\text{-}ONO_2$$

Nitroglycerine was first produced, on the laboratory scale, by the Italian chemist Ascanio Sobrero at the Turin School of Mechanics and Applied Chemistry in 1846 by slowly adding anhydrous glycerine to a mixture of concentrated nitric and sulphuric acids. Nitroglycerine, because of its extreme sensitivity and unpredictability, remained a chemical laboratory curiosity for 17 years until Alfred Nobel was struck with the idea of using the impressive power of the discovery to open up a new era in commercial blasting.

FIGURE 1.
ALFRED NOBEL AGED ABOUT FIFTY

Alfred Nobel was a most remarkable man. Much of his childhood was unsettled and poverty stricken He had virtually no formal school education and never attended University but he became an inventor imbued with a surprising talent and driving force. He had an insatiable intellectual curiosity and a practical genius for finding solutions to problems that other people regarded as insuperable and a determination never to take no for an answer. The success of that philosophy and for which he is internationally acclaimed, is amply demonstrated in his taming of Nitroglycerine.

Nobel realised that the key lay in solving the problem of how to safely and reliably initiate this most unreliable and unpredictable of substances and his solution to that problem, the invention of the Detonator in 1863, has been hailed as the greatest discovery ever made in both the principle and practice of explosives. On it the whole of the modern practice of blasting has been built. The fact that the principle and practice remains virtually unchanged even after 133 years of explosives technological development is, in itself, a tribute to the genius of Alfred Nobel.

The first experimental detonator he produced in a ramshackle out-house in Heleneborg in Sweden, consisted of a safety fuse initiated Blackpowder powder charge contained in a glass tube but, by 1865, the charge had been altered to mercury fulminate in a copper tube and the modern blasting cap was born.

From the very start the demand for the new system of Nobel's detonator and explosive Blasting Oil was dramatic but, unfortunately, accompanied by a whole series of tragic accidents in transport and use. The most devastating to Nobel, however, was the explosion at Heleneborg in experimental production in which his youngest brother Emil was killed. The extreme sensitivity of the Nitroglycerine to impact proved to be too great a hazard for its continued use in the liquid form and Nobel set about finding a way of transporting it safely. After many experiments with different absorbing materials, he finally tried kieselguhr. Kieselguhr consists of the skeletons of the minute sea creature the Diatom and has the ability to absorb many times its own weight of oil. Nobel found that keiselguhr was capable of absorbing four times its own weight of Nitroglycerine giving a red powder which he found was safe to handle. It took the detonation from his mercury fulminate blasting cap to initiate the explosive. When initiated in this way, however, the mixture detonated with virtually all the violence of the liquid Nitroglycerine. Nobel called this mixture Dynamite after the Greek word for power - *dynamis.*

This two part invention - Detonator and Guhr Dynamite - gave to the world the era of the High Explosive and revolutionised mining and civil engineering.

Guhr Dynamite, brilliant invention though it was, suffered from one defect. Containing, as it does 25% of inert material, it does not have quite the same power as the liquid nitroglycerine and some users continued to prefer nitroglycerine in spite of its, by then, well known hazards. Nobel solved this problem with his third major invention - Blasting Gelatine - by substituting Nitro-cellulose (itself explosive) for the inert Kieselguhr. The resultant rubbery waterproof gelatinous explosive is still the most powerful commercial explosive available today and remains the datum against which all other commercial explosives are measured.

During the course of Nobel's lifetime he gathered a cadre of brilliant men around him and continued his experimental work replacing the original Guhr Dynamite and Blasting Gelatine with a whole range of explosives whose effectiveness is governed by two important factors:-

The Power - the ability to do work which, in turn, is dependent on the large quantity of gases produced from the small quantity of solid explosive.

The Velocity of Detonation - the speed with which the reaction occurs and which dictates the shattering effect or Brisance of the explosive. The ability of the high explosive to operate on solid material is however primarily dependent on the shock wave that is invariably produced when the explosive detonates.

These important factors Nobel varied by combining the low performance of the deflagrating explosive of the nitrate/fuel type with the high performance of Nitroglycerine, booting the nitrate/fuel mixtures to detonation with Nitroglycerine. The Velocity is then a function of the amount of Nitroglycerine used. By playing on the permutations and combinations a whole range of different velocities and powers can be obtained.

Nobel was therefore able to design explosives to suit specific tasks. High brisant powerful explosives for blasting hard rock where maximum shattering is required, and more gentle explosives for blasting coal and decorative stone.

The introduction of these new products was followed by a new era in mining and engineering in which many spectacular feats were accomplished which would have been quite impossible with Blackpowder. With huge demands for the growing range of Nitroglycerine based explosives, manufacturing methods improved both in complexity and safety at an amazing rate and Nobel lived to see the introduction of mechanised methods which made a major contribution to process safety as the new industry established itself.

FIGURE 2.
NITRATION - LATE 1880'S

One of the major contributors to the design of these early machines was McRoberts who was appointed by Nobel as one of the first factory managers at Ardeer. His name is particularly associated with Gelatines mixing and catridging which remained in use throughout the Nobel era and whose principles were incorporated in designs in other parts of the world.

FIGURE 3.
McROBERTS GELATINE MIXING MACHINE
LATE 1880'S

FIGURE 4.
McROBERTS GELATINE CATRIDGING MACHINE
1885

Nobel was a pacifist with an abhorrence of dispute and a loathing for war. He had the concept, far ahead of his time, that explosives could be a deterrent to war. He did not, therefore ignore that apparent contradiction between his science and his philosophy but tried to put the one at the service of the other. This explains his technical interest in propellants. The propellants of the time were guilty of barrel fouling and there was considerable interest in a clean propellant. Nobel tackled this problem by extending his Blasting Gelatine concept to propellants by increasing the nitrocellulose content to 50% and adding camphor thereby reducing the velocity and pressures to propellant proportion in the product Ballistite. This fourth major invention did for the propellant industry what Dynamite and Blasting gelatine had done for the high explosives industry.

In the century during which the sophistication of the nitroglycerine high explosive has been available, enormous changes have taken place, and these have been almost entirely due the inventiveness of that one man, Alfred Nobel.

Unlike most inventors Nobel combined technical creativity with commercial flair, both to a very high degree. He had no capital, but that was never a hindrance to Nobel. He borrowed it and within ten years founded an international group of Dynamite Companies in Sweden, Norway, Finland and Germany before turning his attention to the United States, Britain and the rest of Europe.

Of all these, Nobel found Great Britain the most difficult, and that for political reasons. Because of the serious accidents with Blasting oil in different parts of the world the British Government of the day had, in 1869, passed an Act of Parliament which forbade *"the manufacture, import, sale and transport of nitroglycerine and any substance containing it within Great Britain"*. Nobel spent two very frustrating years before he managed to prove the efficacy and safety of Dynamite and got the somewhat reluctant easement of the strict regulations but even then he was unable to obtain permission to establish his business in England. Eventually he turned to Scotland where he found a receptive group of entrepreneurial Glasgow business men headed by John Downie, of Fairfield Shipbuilding.

With Downie's help he looked for and found an isolated location on the West coast of Scotland at the mouth of the Clyde estuary to which he felt the British Government could not possibly object - Ardeer - which he described in a letter to his brother as:-

> *"Everlasting bleak sand dunes with no buildings.*
> *Only rabbits find a little nourishment here; they eat a*
> *substance which, quite unjustifiably goes by the name of*
> *grass. It is a sand desert where the wind always blows,*
> *often howls filling the ears with sand. Between us and*
> *America there is nothing but water, a sea whose mighty*
> *waves are always raging and foaming. Without work the*
> *place would be intolerable."*

Helped by the Scottish corporate principle of Limited Liability, he and Downie set up a Company, The British Dynamite Company, in April 1871 with the rights to work Nobel's Patents.

The first charge of Nitroglycerine was produced at Ardeer on the 13th January 1873. The Company was an immediate success and within two years had cleared its capital outlay and in a further two established a very substantial profit.

Nobel's ambition to establish himself in Great Britain was achieved and he left the company in return for 5% royalty. The Company prospered during the 1880's and developed an impressive overseas trade.

FIGURE 5.
PAINTING OF ARDEER FACTORY
HENRY RUSHBURY R.A.

After Nobel's death in 1896 the company he established at Ardeer in Scotland became Nobel's Explosives Company retaining the Nobel interests in the Canadian, South African and Australian companies and continued to flourish with the demand for explosives during the post 1914-18 war reconstruction period greater than ever before. Finally, in 1926, the chairman Harry McGowan persuaded the related industries of Brunner Mond, British Alkali and British Dyestuffs to come together with Nobel's Explosives as a single chemical corporate entity in the new chemical giant, Imperial Chemical Industries.

The four companies publicly announced their intention to merge on 21 October 1926 and was fully operational by 1 January 1927.

Today use of the sophisticated explosives and devices arising from Alfred Nobel's pioneering work in addition to quarrying, mining and infrastructure civil engineering, has been extended into a bewildering variety of engineering applications such as:-

> Metal forming
> Metal and pipe welding
> Cutting charges
> Furnace tappers
> Perforation charges for oil wells
> Oil and gas exploration
> Underwater blasting
> Passive restraint systems
> Explosive bolts
> Rocketry and guidance on space vehicles

Explosives have made a major contribution in the search for that other vital power source, oil and gas. The use of explosives in the oil industry does not, however, stop at the exploration stage. They are being used to solve major engineering problems associated with the production phase. Once the well has been drilled it is "capped" on the sea bed and the oil drawn off by undersea pipeline. In deep waters laying such a pipeline from the surface causes damaging stresses in the pipe which are dangerous and could cause fractures in the pipe with resultant spillage and pollution. One method which has the potential for eliminating this danger is to explosively weld the pipe on the sea bed as it is being laid.

In retrospect the increasing use of explosives in underwater engineering where there is no oxygen is logical since explosives contain all the oxygen they need for complete reaction in their ingredients and are, therefore, just as effective in producing energy underwater as in air.

Explosives are, therefore, ideally suited to provide high energy in airless conditions. It is for that reason that, as we enter the new millennium, explosives are playing a vital role in the exploration of space. The principle use of explosives in space vehicles is in providing the enormous amount of sustained energy required to lift the space vehicle from the ground and accelerate it to escape velocity. This, today, can only be achieved by the quasi-detonation in the thrust motors of the chemical rocket. These giant motors operate generally on liquid explosive mixtures or, in the case of the smaller guidance systems rockets, on solid propellant explosives.

The dependence on explosives does not, however, stop once the space vehicle has reached its cruising trajectory or orbit. There are numerous other in-flight and landing operations to perform most of which are carried out by explosive devices or gas operated micro-switches. In other words the whole system uses explosives in one form or another in all its stages.

There is cause for philosophical reflection in that, as we enter the new millennium and seriously begin the exploration of the solar system, progress is dependent on the same energy producing chemical reaction that first lifted man from the darkness and ignorance of his Neanderthal beginnings and enabled him to attain the enlightenment which led to civilisation and intellectual dominance of this planet.

The chemical energy of the combustion reaction has served the human race throughout its existence, has sustained its survival and, in the sophistication of the High Explosives of the Nobel Era, has made possible engineering feats which would have been regarded as miraculous even 150 years ago.

Session 2. Legislation and Safety

Explosives Legislation

G. E. Williamson

CHEMICAL AND HAZARDOUS INSTALLATIONS DIVISION, HEALTH AND SAFETY
EXECUTIVE, MAGDALEN HOUSE, STANLEY PRECINCT, BOOTLE, MERSEYSIDE L20 3QZ, UK

1 SUMMARY

The final stage of the programme to replace the Explosives Act 1875 (EA) by modern regulations made under the Health and Safety at Work etc Act 1974 (HSWA) is now firmly in view and will be concerned with manufacture and storage and with licensing provisions largely developed earlier in connection with the establishment of the first dynamite factory at Ardeer around 1871.

New regulations on the manufacture and storage of explosives (MSER) will form part of a wider framework of legislation along with such provisions as the Management of Health and Safety at Work Regulations 1992 (MHSWR) and alongside new regulations to implement a putative EU Directive on the control of major accident hazards (Seveso II) which also refers to explosives inventories of one kind or another.

Reviews have so far suggested it will be appropriate to retain some such licensing provisions as are contained in the 1875 Act but to leave day to day management of health and safety to the sort of risk assessment required by MHSWR. The permissioning provisions of the Seveso II Directive are likely to replace licensing in some cases, at least in safety terms if not for security reasons. The safety reports or major accident prevention policies required by the Directive will clearly satisfy any MHSWR-based assessment processes.

2 INTRODUCTION

By a curious coincidence the final replacement of the Explosives Act 1875 by modern regulations, the subject of this paper to commemorate the centenary of Alfred Nobel's death, is now set to happen in 1998 when international imperatives will finally force the issue. For 1998 will mark the centenary of the death of Vivian Dering Majendie the first of HM Chief Inspectors of Explosives and the one person most instrumental in shaping that Act and bringing it into force.

The last remaining parts of the Explosives Act relate in particular to the control of risks in manufacture and storage and, at the centre of those provisions, the licensing of explosives facilities. Majendie [1-3] was to take important lessons from problems inherent in the licensing arrangements set out in the Gunpowder Acts of 1860, 1861 and 1862, the Dangerous Goods Act of 1866 and the Nitroglycerine Act of 1869, including those which had faced Nobel [4-10] in winning approval to first import dynamite into the United

Kingdom. However, the 1869 Act also provided the vehicle for Majendie to establish the basic principles for the licensing of explosives facilities in discussion and agreement with the British Dynamite Company Ltd the proposed licensees of the factory set up by Nobel in Ayrshire.

The licence to manufacture nitroglycerine and dynamite there at Ardeer was issued on 18 September 1871 and processing commenced in the following year. After the passing of the Explosives Act 1875 the company was granted a certificate (No. 3 of 5 February 1876) authorising the continuance of manufacture under those new provisions and soon after changed its name to become Nobel's Explosives Company, Limited.

Over the next twenty years or so, the range of explosives manufactured at Ardeer was progressively expanded and output became considerably larger than that of any other factory in the world. The site had grown at least four times as large as it was in 1876 to around 500 acres. The number of work-people and the number of buildings they occupied had both increased by rather more, from 45 to around 450 persons and from 38 to 315 buildings provided for by a licence which had been added to or otherwise amended as necessary to keep pace with that ongoing innovation and expansion.

The progressive development of the licence to reflect new knowledge and technological change or to incorporate the lessons learned from a number of incidents at the site are well set out in contemporary Reports [11-14]. But those reports also well reflect the exceptional standards achieved by site managers at Ardeer and their readiness to adopt additional precautions when suggested. A letter addressed by Majendie to all of the people at "Fac. 3 Ayr" on 21 November 1896, just three weeks before the death of Nobel, shows clearly how much impressed he had been during an inspection visit, "not merely with the order and discipline and generally strict observance of the regulations which prevailed throughout the whole of that extensive establishment, but with the numerous and valuable devices which had been adopted, as the result of accumulated experience and observation, to diminish risk and to localise, so far as may be, the effects of an accident, should one unfortunately occur".

Majendie's suggestions for a new Act show that he would have appreciated the distinction we can now draw more clearly than he could between the localisation of the effects of an accident with the reduction in risks that wins, and the focus that can then be brought to bear on the proper control of the residual risks inherent in each and every single operation. His suggestions find echoes in modern practice as set out by Robens [15] in 1972 and as re-stated since by the Health and Safety Commission [16].

The Explosives Act had proved itself successful over that first period of twenty or so years up to the deaths of Nobel and Majendie. Many new explosives had been brought under its cognisance and the industry had been free to expand in a way that could not have been foreseen when the Act was first promulgated. It was adaptable to the requirements of those new days. What now remains is adaptable to the explosives industry as it now reforms. Experience lends support. Current reviews suggest it will be appropriate to retain some such licensing provisions.

3 ORIGINS

In the present context it is worthwhile to look again and in some more detail at the main issues addressed by Majendie in his reviews of the Gunpowder and Nitroglycerine Acts, the origins of the Explosives Act 1875.

3.1 Gunpowder Acts 1860-62

Inspection of the ammunition, firework, cap and mercury fulminate factories in and around Birmingham following a number of disastrous explosions there towards the end of 1870, and including one in which 53 people were killed, showed all to be unsatisfactory in terms of both their disregard for the Acts and any other precautions essential for safety. And much the same picture emerged from a wider inspection of factories and magazines across the United Kingdom.

The Gunpowder Acts were found deficient in many respects but not least because they were too stringent in some, not stringent enough in others, were too inflexible or were insufficiently comprehensive. Their failure was further underlined by the lack of systematic inspection by competent officers.

Majendie raised six main issues and with them a series of recommendations for a new Act. The first issue concerned the need to make provision for such explosives as guncotton not apparently in scope of the old legislation and any new explosives. Other issues concerned:

3.1.1 The Prevention of Accidents. The Acts had been almost silent on the need to limit the probability of accidents though provision had been made, if inadequately, to protect the public from the effects of any explosion. Remedies resolved into the General and Special Rules provided for in the 1875 Act but which would nowadays fall to legislation on the management of health and safety or to approved codes of practice and other guidance. Modern day legislation also addresses questions of competence which Majendie had posed but could not resolve by certification of managers.

The remedy to inadequate licensing was to allow licences for factories and magazines to be granted at the local level as before but only against the certificate of a competent officer. This device had the further benefit that expert judgement could be brought to bear on the variation of quantity distance relationships, for example to take account of the risks involved or say when topographical features may have effect to limit potential consequences.

3.1.2 The Protection of the Public. It was considered impossible to prevent all accidents and not least because there could be no guarantee against the occasional relaxation of normal precautions. Experience, such as that at Stowmarket in 1871, had also suggested that however unlikely an explosion might appear it remained important to treat all explosives as liable, under some condition or another, to be exploded - even wet guncotton.

In order to give the better protection to the public that implied was necessary, the large or unlimited keeping of explosives in private houses or in retail premises was stopped and inventories switched instead to licensed places to which quantity-distance relationships could be applied. Much work was necessary to establish a more comprehensive and rational quantity distance system, with fixed relationships applied at the smaller places under local authority control and with the more flexible approach seen as necessary for factories and magazines.

3.1.3 The Retail Trade. The mere fact of selling powder constituted a dealer, and the mere fact of dealing entitled that person to keep 200 lbs anywhere and anyhow, and without any supervision. There were many explosions and many were killed, carelessness had seemed endemic. The remedy was to apply the systems of licensing and registration and with them the fixed rules detailed in Order of Council. The fixed rules were carefully targeted to take account of the different natures of explosives, degrees of risk and modes of keeping.

3.1.4   Inelasticity. The Acts had been shown to be inadequate in many respects and not least in Majendie's view because they included no provision whereby the growing requirements of the industry could be satisfied or regulated. He saw discretionary powers in matters of detail as a potential solution, quoting powers already contained in the Nitroglycerine Act 1869 to specify and change the provisions of the licences it required.

The idea that licensing possessed an inherent power of adaptation to varying conditions that Majendie likened to an "elastic fence which can be stretched round any new ground which may become occupied by the advances of science" would have been strengthened by the successful licensing of the British Dynamite Company Ltd at Ardeer and the ability to amend those provisions as knowledge and experience accumulated.

However it seems unlikely that Nobel would have been entirely in agreement following the difficulties he had faced in attempting to import dynamite for the first time into this country. And indeed Majendie was soon to amend his ideas on the matter.

3.2   Nitroglycerine Act 1869

This Act required any person who imported, exported, manufactured, sold, carried or otherwise disposed of or had in their possession any quantity of nitroglycerine, whether or not in a composition, to be licensed by the Secretary of State. Penalties were severe.

3.2.1   Importation of Dynamite. Though not strictly relevant to the main theme of this paper, the problems faced by Nobel in winning approval to import dynamite do serve to underline the significance of Majendie's work in licensing the factory then soon set up at Ardeer and in establishing principles adopted by every major country across the world.

Suffice it to say that a War Office Committee on Explosives [4] found that the "highly dangerous character of nitroglycerine was not sufficiently modified by conversion to dynamite to warrant any exception from restrictions on transport, storage and use". The Committee believed those restrictions further justified because in its view both nitroglycerine and dynamite could be produced on-site for immediate use in simple and safe manufacturing processes. The Home Department was not impressed and neither was Nobel.

Letters were exchanged [5-8] , and envoys despatched to look into reports and testimonials held by Nobel and to witness tests all intended to show that dynamite had been used to advantage for several years in many countries and without accident. The safety of guncotton was unfavourably compared with that of dynamite and motives questioned before the very many eminent politicians, scientists and officials involved could move towards the grant of general licences for storage and carriage [9,10.]

Agreement on the terms of the general licence proved no less easy or time-consuming. Nobel's own assessment of the risks involved in storage suggested to him that large tonnages of dynamite could be kept safely in wooden buildings anywhere within municipal boundaries. The Secretary of State would not have objected to the storage of small quantities of dynamite for use at quarries and mines but was adamant that if large quantities were to be imported and kept then proper sites must be found and plans of proper buildings submitted.

3.2.2 Manufacture of Dynamite. The factory of the British Dynamite Company at Ardeer was completed in the latter part of 1872 and the large scale manufacture of dynamite and its corresponding distribution and use was quickly established. Provision had been made in time to meet anticipated demands for licences and more than 400 were issued over the next year without any great difficulty. But several problems emerged, the railways for instance were unwilling to accept a licence and refused carriage and the effect of the central concentration of the work and responsibility for licensing could not be resolved without new powers.

Repeal of the Nitroglycerine Act allowed the inclusion of nitroglycerine explosives in the new Explosives Act along with the rest.

4 THE FRAMEWORK FOR REVIEW

A modern framework of explosives legislation is already largely in place. Figure 1 shows the pre-eminence of the Health and Safety at Work etc Act 1974 (HSWA), some examples of the sort of subordinate legislation which applies generally across industry, including the explosives sector, and the specialist regulations on explosives which have already replaced large parts of the Explosives Act 1875 (EA).

The remaining provisions of the Explosives Act along with some 45 subsidiary Orders, all relevant statutory provisions of the Health and Safety at Work Act, are set in broad terms to be replaced by the new regulations on the manufacture and storage of explosives, given that and the corresponding acronym (MSER) as working titles.

In completing that transformation, reviews already suggest that particular reference will need to be made to the general regulations on the management of health and safety at work (MHSWR) and the regulations which will implement before the end of 1998 an EU Directive on the control of major accident hazards (Seveso II) and which will embrace explosives as much as other dangerous substances. The Seveso II and MSER provisions will be inter-dependent and because of that will need to be made at the same time.

4.1   Review Processes

The review processes and the key outputs in terms of goal setting regulations, approved codes of practice where appropriate and clear, targeted guidance have been well described elsewhere in two major reports [15, 16]. The focus now placed on the needs of small business will clearly be of relevance to the explosives industry as it continues to rationalise and reshape itself.

The aim, enshrined in law, will be to maintain existing standards of health and safety and where possible improve them in so far as cost-benefit balances will allow. Maintenance or improvement will be judged in terms of the overall framework of legislation not just as a function of the specific manufacturing and storage regulations in view. With deregulation in the background, the aim cannot be to impose higher standards *per se*.

The starting point is to consider in full consultation with all of the social partners what solution appears right in principle given the framework of law already in place, and then to consider the potential impact of other influences like Seveso II in order to determine how the final result might be achieved.

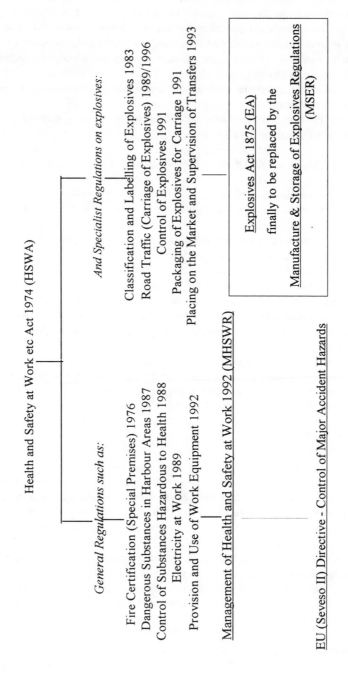

Figure 1 - The Framework of Explosives Legislation

In bringing so much together care will have to be taken to ensure no gaps are left, but particular emphasis will also have to be placed on the various interfaces to ensure there will be no overlapping duties with similar purposes applying to the same operation. Provisions for example aimed at the prevention of accidents in MHSWR may need to replace provisions in EA which have the same purpose and which might otherwise have been carried forward into MSER.

Review processes clearly have their own acronym-based language and the need for that is further emphasised when considering how general aims and objectives apply at the interface with EU Directives. The United Kingdom will be open to challenge if EU Directives are not implemented in full. In some cases, like Seveso II, the EU measures may be extended by national provisions, in others they must stand as completed. Provisions in MSER may neither overlap nor replace provisions in Seveso II which have the same purpose, but may extend those measures in other cases when the Directive is silent on a matter which needs to be applied across the whole of the explosives industry, such as on security. However, the need to avoid the imposition of higher standards and with them added burdens on industry makes it necessary to ensure that provisions unique to Seveso II such as those on environmental protection are not applied more widely.

4.2   Explosives Act - Manufacture and Storage

The potential for overlap of the kind just described is likely to be most real in relation to the licensing by HSE of explosives factories and magazines, and possibly in relation to the various provisions for general or special rules set out in the Act and subordinate Orders.

4.2.1   Licensing. The draft licence for a proposed factory or magazine is first subject to technical assessment and approval by HSE and then to the assent of the local authority in consultation with the public before being confirmed and issued by HSE once satisfied the place is sufficiently completed and ready for operation.

The licence must include a map of the relevant area, a plan of the site and should specify such matters as the construction of buildings and mounds, the nature of the processes and places concerned, the quantities of explosives in each building, the corresponding distances to protected works off-site, the maximum number of persons in each building and any special terms suggested by the applicant or by HSE. In modern practice, the sort of detail which would be liable to frequent amendment is omitted as far as possible.

4.2.2   General and Special Rules. General Rules are set out in the Act to cover the sort of basic safety provisions more likely nowadays to be found in an approved code of practice or other guidance. Most are concerned with potential sources of ignition.

The occupier of a licensed factory or magazine is also required to make Special Rules which are to cover the regulation of people on site and their safety, matters addressed to some extent by the general provisions of HSWA or otherwise by MHSWR, approved codes and guidance.

4.3   Management of Health and Safety

MHSWR requires an employer to make a suitable and sufficient assessment of risks to people in order to identify the counter-measures required to comply with health and safety legislation. Where five or more employees are involved, the significant findings are to be recorded and made available to an HSE Inspector on request.

Initial findings suggest a slight tightening of those provisions to require that where explosives are involved the risk assessment should always be available in writing and provided irrespective of size of the undertaking. Significant changes to that risk assessment may also be made notifiable.

4.4   Seveso II Directive

This directive will replace an earlier one which was implemented by the Control of Industrial Major Accident Hazards Regulations 1984. Those regulations did not apply to any of the facilities subject to the Explosives Act 1875 but the position now changes, the replacement regulations on the control of major accident hazards which implement Seveso II will. Current projections suggest that of the 250 or so factories or magazines licensed by HSE some 20 will be subject to the so-called "Top Tier" requirements and a further 20 to the less stringent "Lower Tier" provisions.

Seveso II applies to establishments where dangerous substances, and including explosives also in articles, are present in excess of specified quantities. Limits are set on aggregated quantities of all classes of dangerous substances present and may be dependent as well on the sensitiveness or toxicity of an explosive.

The directive does not apply to military establishments, mineral extraction sites like mines or quarries, or to transport and related transit storage.

Provision is made for emergency planning, land-use planning, the provision of information to the public, accident reporting and inspection but those relating to the notification of establishments, the preparation of a major accident prevention policy (MAPP) or submission of a "safety report", according to tier, hold most interest in the context of this paper.

4.4.1 Lower Tier - MAPP. All establishments subject to Seveso II and including those with the smaller qualifying inventories designated as "lower tier" will have to be notified and will be required to draw up a document which sets out their major accident prevention policy. The document, which must be made available to HSE, will cover the operator's overall aims and principles of action with respect to the control of major accident hazards and the safety management system for determining and implementing the MAPP.

4.4.2 Top Tier - Safety Report. In addition to the duties just outlined, the occupier of a "top tier" site will be required to submit a safety report to HSE at specified times or intervals, to demonstrate that major accident hazards have been identified and that necessary measures have been taken to prevent such accidents and to limit the consequences for man and the environment. The report must address reliability issues, internal emergency planning and land use planning. Descriptions of the site and its environment, main activities, processes and dangerous substances inventories must also be included along with a risk assessment.

HSE must communicate the results of its assessment of the safety report to the operators, for example before a new installation is constructed or operated and may in some circumstances prohibit its bringing into use. The assessment of the safety report is a form of prior approval "permissioning" in the same sense as the licence for factories and magazines required by the Explosives Act 1875.

5 PROPOSALS

The challenge is to make new regulations on the manufacture and storage of explosives which will form a coherent part of the overall framework of health and safety legislation and which in that context will maintain the standards established by the Explosives Act 1875. The solution that appears right in principle at this time is to set out a code of general duties that would apply right across the explosives field and to maintain some form of prior approval permissioning alongside provision for risk assessment.

Those three broad strands could be provided by separate Seveso II Regulations and regulations on the manufacture and storage of explosives and by MHSWR, if linked together and supported by approved codes of practice and guidance. General duties apart, MSER would make a distinction between places controlled by local authorities and which as now would be subject to fixed rules, and other places which would be licensed by HSE in accord with published guidelines.

5.1 General Provisions

MSER will need to define the explosives in scope and to which the general duties will apply whether or not the Seveso II Regulations also have force. The definition is likely to have much the same effect as that in EA though classification will owe more to the United Nations system embraced in modern explosives legislation than to Order in Council No. 1 which can now be dropped.

The general duties will need to address such matters as the places where explosives may legally be kept, the security of explosives and the prevention of unauthorised access or for example the prevention of accidents by fire or explosion. Some matters may need to be carried forward from the General or Special Rules and so in that way allow their repeal. It may also prove appropriate to deal with some points now covered by licensing by way of a general provision if not left to MHSWR and approved code of practice, an option also in relation to the Rules.

Provision will need to be made for site mixing operations for immediate use.

5.2 Permissioning Regimes

These are best described by reference to the top tier and lower tier sites to be subject to the Seveso II Regulations, and the places subject to fixed rules or licensing by HSE under MSER.

5.2.1 Seveso Top Tier. These sites will be subject to the safety reporting procedures described above and no further permissioning, such as by licence, will be required. Guidance on the assessment of risks from explosives and on the consequences of explosions can be provided, as can the usual kind of Quantity-Distance tables which provide guidelines for licensing.

5.2.2 Seveso Lower Tier. These places will not be subject to prior approval under Seveso II Regulations and will need to be licensed by HSE under MSER. Suitable links and support will be provided by approved codes and other guidance.

5.2.3 MSER Fixed Rule Sites. Though not strictly a permissioning regime, licences will only be granted and remain valid if the fixed rule conditions are met in full. The regime will apply to the licensed stores and registered premises now subject to local authority control under EA 1875. However that distinction will be dropped in favour of a simple licensing provision which will apply across three broad sectors covering fireworks, ammunition and all other explosives.

Sub-division into those sectors could offer several advantages not least those of better targeting and simplification. It resolves the problem that mixing rules cannot be based on the classification system set out in the Explosives Act. It allows the development of more flexible quantity limits and corresponding distances to protected works which will in turn allow the most economical use of available space and particularly at places now registered. It allows enforcement to be targeted in the most appropriate way, for example by making trading standards officers responsible for fireworks or the police for ammunition and other explosives. Or again, it allows a larger range of operations to be sanctioned at that level, such as the fusing of lancework or the preparation of additional explosives like ANFO for immediate use.

Further flexibility could derive from the use in some circumstances of mobile stores if security requirements can be made more explicit.

5.2.4 Licensing under MSER by HSE. Again there appears considerable scope for simplification. Licensing could merely relate to places not subject to Seveso II top tier requirements and not subject to the fixed rule regime just described. Removal of the distinction between factories and magazines avoids further difficult and uncertain definition, allows a site to be designed and licensed for purpose. HSE would in essence licence just those places it was not for some reason or another appropriate to place under the fixed rule regime.

Licensing as under EA and described at 4.2.1 above remains the model for the provisions in view though there will need to be further consideration of the elements which might be left now, such as day to day safety management issues, to either a general duty or to the risk assessment under MHSWR.

Initial thinking suggests there could be merit in a two part licence; a part concerned with the more permanent features of a site like its positioning and layout, the construction of buildings and security arrangements as well as quantity limits and distances to protected works, and a part which referred to the use of a building whether for processing or keeping. The first part contains the information most relevant to assent procedures and to off-site emergency or land use planning arrangements. The second part refers to matters which might be subject to more frequent change and which need sanction by HSE alone. The latter part could provide useful pegs on which to hang the risk assessment left to MHSWR. New guidelines are in view.

5.3   Risk Assessment

Risk assessment will in essence be left to MHSWR which applies in any case across the board. But it is recognised that the Major Accident Prevention Policy required for all places subject to Seveso II and the safety reporting provisions of the same regulations will equally apply and will strengthen those requirements. Approved codes of practice and other guidance can provide support to MHSWR as applied to places subject to MSER, just as they can to other relevant legislation which generally applies such as the Provision and use of Work Equipment Regulations 1992.

6 CONCLUSIONS

Much work remains to be done before the end of 1998 to mould and finally shape the regulations which will replace the last remaining parts of the Explosives Act 1875. In that sense it is yet too early to draw conclusions but it seems likely that basic principles established a hundred or so years ago and well shown to be successful will nonetheless be carried forward in the new provisions.

References

1. Vivian Dering Majendie, Reports on the necessity for the amendment of the law relating to gunpowder and other explosives. Presented to both Houses of Parliament by command of her Majesty, 1874.
2. Annual Report of HM Inspectors of Explosives for 1875.
3. Annual Report of HM Inspectors of Explosives for 1898.
4. Report of Committee on Explosives of War Office, 8 September 1869 (HO 45-OS 8245 11).
5. Alfred Nobel, letter of 27 September 1869 to H A Bruce Secretary of State, Home Department (HO 45-OS 8245 12).
6. Mr Ward, letter of 22 October 1869 to the Earl of Clarendon KG. (HO 45-OS 8245 15).
7. Petition on behalf of Messrs Alfred Nobel and Co of Hamburg, 24 November 1869, to the Right Hon. Henry Bruce, Her Majesty's Secretary of State, Home Department (HO 45-OS 8245 16).
8. Alfred Nobel, letter of 27 December 1869 to H A Bruce, Secretary of State, Home Department (HO 45-OS 8245 19).
9. File Note, 3 February 1870 (HO 45 - OS 8245 22).
10. Alfred Nobel, letter of 28 March 1870 to H A Bruce, Secretary of State, Home Department (HO 45-OS 8245 31).
11. Captain J P Cundill, Report on the explosion which occurred in the factory of Nobel's Explosives Company (Limited), at Ardeer near Stevenston, in the County of Ayr, on 8th June 1882.
12. Colonel A Ford, Report on the explosion which occurred in a cartridge hut at the factory of Nobel's Explosives Company (Limited) at Ardeer, Ayrshire on the 8th May 1884.
13. Captain J H Thomson, Report on the explosion which occurred at the Dynamite factory of Nobel's Explosives Company (Limited) at Ardeer, near Stevenston, in the County of Ayr, on the 5th January 1895.
14. Colonel A Ford, Report on the explosion of Nitro-Glycerine which occurred in one of the final washing-houses of the factory of Nobel's Explosives Company, Limited at Ardeer, near Stevenston, Ayrshire, on the 24th February 1897.
15. Safety and Health at Work, Report of the Robens Committee 1970-1972, Cmnd 5034, July 1972.
16. Health and Safety Commission, Review of Health and Safety Regulation, May 1994.

The CEN Harmonisation Programme

F. M. Murray

SECRETARY GENERAL, FEDERATION OF EUROPEAN EXPLOSIVES MANUFACTURERS,
KIRKFAULD, KILMAURS, AYRSHIRE, UK
CHAIRMAN, CEN/TC321

SUMMARY

In the middle of the 1980s, the 1992 Programme proposed a free market where goods, services and capital could move freely from one member state to another within the European Union without let or hindrance.

In the case of explosives, where member states each had a long history of testing and approval by national authorities before products are placed on the market, the European Commission used a "New Approach" Directive which would control all explosives which were intended to be sold in Europe.

This Directive is known as 93/15/EEC and has two main thrusts:

1 The harmonisation of transfer documentation within the European Union; and

2 The essential safety requirements of an explosives product before it can be placed on the market.

In addition, the Directive visualised setting up test houses, notified by National Governments to the Commission ("Notified Bodies") which would test the explosives and ensure they meet the essential safety requirements of the Directive.

Having satisfied itself that the explosive product meets the essential safety requirement of the Directive, the Notified Body then issues a "CE" mark for the product which allows it to be placed on the market anywhere within the European Union.

With regard to timing, the Directive requires that Member States adopt the "Harmonisation of Transfer" documents by September 1993, and that all products placed on the market after 1st January 2003 should have a CE mark.

The European Commission then gave a mandate to the Comité Européen de Normalisation (CEN) to create a series of harmonised tests and criteria which would establish if an explosive product met the essential safety requirements of the Directive or not.

In turn, CEN appointed a technical committee, to be known as Technical Committee 321 (CEN/TC321) which would carry out the mandate of the Commission and describe standard tests and criteria for explosives which are to be placed in the market within the European Union.

This paper describes the work of CEN/TC321, its composition, its modus operandi and progress to date.

In summary, progress is such that harmonised tests and criteria for explosives approval for placing in the market are expected to be complete by the year 2001.

1 INTRODUCTION

The Treaty which was intended to establish the European Economic Community proposed in Article 8A that the internal market of the European Union must be established not later than 31st December 1992. This was known as the Single European Act (or The 1992 Programme). The intention of this programme is that the internal market of the European Union is to comprise an area without internal frontiers in which the free movement of goods, persons, services and capital is ensured. Of course, the free movement of goods presupposes, particularly with respect to explosives, that certain basic conditions are fulfilled: in particular the harmonisation of laws concerning the placing of explosives on the market.

However, because of the long history of the development of explosives, each of the member states within the European Union has produced detailed national regulations, mainly in respect of safety and security and, consequently, because of the differences between the regulations in member states these act as a barrier to cross-border trade.

To solve this problem the European Commission produced a Directive specifically designed to allow the free cross-border movement of explosives for civil uses.

This was a "New Approach" directive which recognised that European standards were not harmonised and recognised the European Committee for Standardisation (Comité Européen de Normalisation - CEN) as a body capable of producing harmonised standards for the European Union.

2 DIRECTIVE 93/15/EEC- PLACING EXPLOSIVES ON THE MARKET

This Directive was adopted by the Commission in April 1993 and has two main legislative thrusts:

1   The harmonisation of transfer documentation within the European Union; and

2   The essential safety requirements an explosive product must meet before it can be placed on the market.

2.1 Harmonisation of Transfer Documents

The basic concept behind this legislative thrust is that in one single open common market, with the free movement of goods, there is no such thing as a "domestic transfer" or an "international transfer" of explosives. Therefore the documentation process within a member state should be the same as that from one member state to another.

The objective here is to ensure that member states do not use a bureaucratic process to hinder the free movement of explosive goods. Basically, the Directive requires that the approval to transfer explosives shall be obtained by the consignee from "a recipient competent authority". This "competent authority" has a duty to ensure that the consignee is authorised to acquire explosives and that he is in possession of the necessary licences etc.

The competent authority shall then issue a document which will accompany the explosives on its delivery specifically allowing the transfer of the products.

The Directive requires that this piece of legislation be enacted by the member states by September 1993.

A specific and peculiar problem, however, arises with explosives. Because of the nature of the products most users are reluctant to store them at the point of usage. This is not only because of the inconvenience of needing custom built magazines but also because of the need for appropriate security measures, which are required to prevent theft and subsequent abuse of explosives.

Consequently, throughout Europe the explosives industry has developed a very fast delivery service. In many instances, this is a "next day" delivery service where the consignee phones his order to the supplier for the precise amount required in a blast and expects delivery on the following day. If this is done after the consumption figures for the blast have been calculated, then the need to store explosives at the point of usage is minimal and little or no explosives are stored on site.

Given this situation, the requirements of the Directive for harmonisation of transfer documents across Europe become extremely difficult. Clearly, for a competent body to authorise the transfer of explosives on a day-by-day basis is almost impossible in a "domestic transfer" situation.

By contrast, in an "international transfer" situation next day delivery is not normal and under these circumstances, therefore, it would be just possible for competent bodies to issue transfer documents.

However, this does not match with the requirements of the Directive for harmonisation of documentation transfer, between "domestic" and "international" transfers.

In the UK, the problem has been partially solved by the "season ticket" approach wherein the competent body (The Health & Safety Executive) issues transfer documentation to consignees to purchase certain explosive products identified by the UN Orange Book code number for a period of time (usually a year).

However, this approach while satisfying the requirements on a domestic basis does not meet the general requirements of the Directive for a uniform method within the European Union.

In addition, the transfer of ammunition for civil purposes, which is done on a much more sporadic basis than the delivery of explosives, causes an immense bureaucratic burden on the "competent authority".

Although the European Directive calls for national state enactment of this piece of legislation by September 1993, to date there has been no solution to the technical difficulties, and therefore this part of the legislation is more honoured in the breach than in compliance.

2.2 Essential Safety Requirements

The Directive also requires that members shall take the necessary steps to ensure that all civil explosives products placed on the market meet the essential safety requirements as described in its annex I. In addition, the Directive requires that all civil explosives placed on the market will be provided with a CE mark.

However, with regard to timing the Directive goes on to propose that CE marking shall not be necessary for all products until 1st January 2003. However, during the period up to 31st December 2002, member states will be permitted to place on the market explosives complying with national regulations in force in their territory before that time.

The Directive visualises an "interim period" between 1st January 1995 and 1st January 2003 during which time CE marks may be issued even although harmonised standards are not yet in existence.

Thus, the Directive requires the following:

1   All explosives products placed on the market within the European Union after 1st January 2003 must have a CE mark.

2   Between 1st January 1995 and 1st January 2003, explosives manufacturers may continue to supply products within "their markets" provided the products meet the requirements of the national authorities.

3   Explosives manufacturers may "export" their products to another member state of the Union provided the products meet the national requirements of that country.

4   Even in the absence of harmonised standards, manufacturers may apply to notified bodies for CE marks between 1st January 1995 and 1st January 2003. The notified bodies under these circumstances must satisfy themselves that the products meet the essential safety requirements of the Directive.

2.3 Notified Bodies

Directive 93/15/EEC visualises that member states shall appoint test houses which are qualified to test explosives whose task it would be to examine products and to award CE marks.

The member states would notify the European Commission of these bodies and hence they become known in shorthand as "Notified Bodies".

To date only four bodies have been notified to the European Commission by member states and these are:

UK   - The Health and Safety Laboratories at Harpur Hill, Buxton.

France   - Ineris at Creil

Spain   - LOM at Madrid

Sweden   - SP, Stockholm - detonators only

Other member states anticipate notifying the Commission of test houses, but for various bureaucratic reasons are not yet formally on the list of notified bodies and they include BAM, Germany; TNP, Holland; Norwegian Testing Authority, Norway; Austrian Testing Authority, Austria.

In the absence of harmonised standards for testing and identifying the characteristics of explosives within the European Union, the Notified Body must first:

1   Ensure that the product submitted for test meets the essential safety requirements of the Directive. In the interim period between 1st January 1995 and 2003 and in the absence of harmonised standards, Notified Bodies have a certain amount of freedom to decide what tests and criteria they will use to ensure that the products submitted meet the essential safety requirements of the Directive.

The programme for producing harmonised tests is scheduled to be complete and agreed upon by 1st January 2003. After that date, the use of harmonised standards is still not mandatory. However, if a Notified Body awards a CE mark to a product which is shown to clearly fail the harmonised tests then the award of that CE mark is open to challenge by other member states within the Union and by other Notified Bodies.

This primary examination of an explosive product is described in Annex II of the Directive as "Module B:EC Type Examination". This describes the part of the procedure by which a Notified Body ascertains and attests that an example representative of the product envisaged meets the relative provisions of the Directive.

After the Notified Body has satisfied itself that the product is suitable and does meet the essential safety requirements of the Directive, the next step is for the Notified Body to ensure that the sample presented is representative of the product which is supplied for normal usage.

In this the manufacturer must choose one of four different methods which the Notified Body must then apply.

These are:

2.3.1 Module C : Conformity to Type.   This is a procedure whereby the Notified Body chosen by the manufacturer must perform examinations of the product at random intervals. A suitable sample of the finished product taken on the spot by the Notified Body is examined and appropriate tests carried out; or:-

2.3.2 Module D : Production Quality Assurance. This   is   a   Production Assurance system in which the manufacturer must operate an approved quality system for production, which must ensure conformity of the explosive products with the type as tested under Module B; or:-

2.3.3 Module E : Product Quality Assurance.   This module describes the procedure whereby the manufacturer operates a quality system whereby each explosive product is examined and appropriate tests as defined in the relative standards or equivalent tests and carried out in order to verify its conformity with the relevant requirements of the Directive; or:-

2.3.4 Module F : Product Verification.   In this the manufacturer takes all measures necessary to ensure that the manufacturing process ensures conformity of the explosive type with the type as described in Module B.

Thus the Notified Body in summary, has the duty to first of all examine the product and then to examine the manufacturer's system of ensuring that the product supplied to the market place conforms with the product submitted to test.

This procedure is summarised in Appendix I.

2.4 The Comité Européen de Normalisation (CEN)

CEN is an international association set up to manage the co-operation amongst the national standard bodies (e.g. BSI, DIN, AFNOR, AENOR, etc.)

The first step of CEN was to pull together a working group of delegates from each of the member states in an attempt to write a draft work programme of potential standards.

Subsequently, CEN appointed a technical committee known as CEN/TC321 with a mandate from the European Commission to produce a series of harmonised tests and criteria to match the essential safety requirements of Directive 93/15/EEC.

CEN then appointed the Spanish Standardisation Institute (Association Espanol de Normalisation (AENOR) as the secretariat responsible for the production of the standards and they in turn appointed F M Murray as chairman of the technical committee.

2.5 CEN/TC321

CEN/TC321 comprises delegations from each of the member states of CEN. It should be noted that this is eighteen rather then the fifteen members of the EU, because of the accepted presence of associate members (Norway, Iceland, Switzerland). Each member is allowed a delegation of no more than three individuals, thus a theoretical total of fifty-four could attend a plenary session of TC321.

In practice, membership tends to be from those countries with a specific interest in explosive standards and plenary sessions are usually attended by thirty to thirty-five members.

The real work of producing the standards is done in working groups and there are five of these divided according to the grouping in the Directive as follows:

Working Group 1   Nomenclature and Co-ordinating Committee
Working Group 2   Explosives
Working Group 3   Detonating Cord
Working Group 4   Detonators
Working Group 5   Propellants.

2.6 TC321 Working Groups

The working groups of Technical Committee 321 are composed of specialists in the field of explosives, detonators, detonating cord and propellants.

Membership of the working group is proposed by the standardisation institutions and the experts may come from manufacturing industry, national laboratories, notified bodies, or national authorities.

In the case of TC321 working groups, 60 to 70 per cent of representatives come from the manufacturing industry, and the remainder from National Authorities, or National Laboratories.

The convenorship of the working groups is decided at the plenary sessions of the technical committee and are allocated to CEN members who in turn appoint a specific individual.

The remit of the working groups is to produce standards or part standards for each one of the work items which appears in the mandate given to CEN/TC321 by CEN.

Appendix II shows the complete work programme of the technical committee and clearly identifies the responsibilities of each one of the working groups.

When a working group is satisfied that it has reached a first draft it is then submitted for approval to the Technical Committee. Once this is agreed by the Technical Committee the draft standard is then translated into the three official languages of CEN - English/French/German and finally the draft standard is submitted to the members of CEN for approval.

Approval, in the final analysis, is by qualified majority voting of each of the member states, but in the reiterative process of submission of draft standards to the TC and for internal enquiry within the membership of CEN, once a standard goes out for final approval, it is rarely rejected at the voting stage.

2.7 Modus Operandi of TC 321 Working Groups

Each working group has approximately fourteen to fifteen experts in their own subject matter as members.

From the attached annex of the work programme and mandate from CEN it can be seen that there are fifty-five work items or standards.

Each one of these work items is designed to meet one or more of the essential safety requirements of the Directive.

As previously indicated, in parallel with the format of the Directive, the working groups are divided into explosives, detonators, detonating cord and propellants. In addition, there is a co-ordinating committee known as Working Group 1 which comprises the convenors of each one of the four main working groups, plus a single representative from each one of the other members of CEN.

For example, the explosives working group - Working Group 2 has as its remit work items 4 to 18.

Wherever possible, the test methods have coincided with the test methods described in the United Nations "Orange Book - Recommendations on the Transport of Dangerous Goods - Tests and Criteria - Second Edition".

In addition, it can be seen from an examination of the work programme that the requirements or criteria for each standard have been separated, at least during the work programme, from the description of the test methods. Thus, for example, Working Group 2 in work item 6 describes a test method for the determination of friction sensitivity of an explosive product but, at this stage, has not discussed the pass/fail level.

Of course, in the final standard the criteria for success or failure (requirements) must be an integral part of the standard, but during the work programme for the sake of progress and efficiency in the initial stages the working groups have confined themselves to description of test methods only.

For each one of the fifty-five work items in the mandate from CEN, a project leader (or a work item leader) has been appointed with the purpose of preparing the initial draft. This is then submitted to the working group for refinement and, after consensus has been reached, a first draft is prepared.

This first draft is submitted to the plenary of the technical committee for general approval. Any comments are taken from the technical committee at this stage and incorporated, again by consensus, into an amended draft. The final draft is then submitted to an editing committee for translation into French and German.

It should be emphasised that the working language of the committee has from the very inception been English without simultaneous translations into the formal requirement of the three languages of CEN.

With the approval of the French and German delegates the working language is first of all English, but of course the formal approval of all documents has to be made in the three languages of CEN - English, French and German.

Once the standard has been satisfactorily translated into the three official languages, it is them submitted to CEN who launch a formal vote from its members.

2.8 Summary of Progress to Date

One of the difficulties of standardisation work within a highly technical and critical area such as explosives is that explosives experts have very little past history or knowledge of standardisation processes and, likewise, standardisation specialists know little or nothing of the explosives industry.

In order to establish some momentum in the process it was agreed that each of the standards would be divided into the category of;

- easy
- medium
- difficult.

Naturally, the easy items were tackled first in order to allow explosives specialists to become better acquainted with standardisation work and standardisation specialists become better acquainted with explosives technology. This tactic has been broadly successful and, at the time of writing, approximately 80 per cent of the fifty-five work items have reached the first draft stage.

Since this work is not funded by the European Commission or by CEN, but rather by the individuals supporting it (and their employers) it is not deemed wise to have a meeting frequency of more than twice per annum.

In practice, what happens is that Week 12 and Week 38 of each year are pencilled in as a "week of meetings" of the working groups of CEN/TC321. Meetings are scheduled so that explosives and detonators run in parallel and detonating cord and propellants run in parallel. The steering committee meeting is held midweek and this general arrangement allows for a good deal of interaction between the various personalities involved.

The main work of the explosives experts will be completed after the first draft stage, and thereafter the main thrust of the work tends to be bureaucratic and linguistic. Moving the standards into the three official languages of CEN and ensuring simultaneously that there are no major mis-translations is the major task which lies ahead.

The Work Programme indicates that all standards will be completed by the year 2001 and at the current rate of progress there is no reason to doubt this.

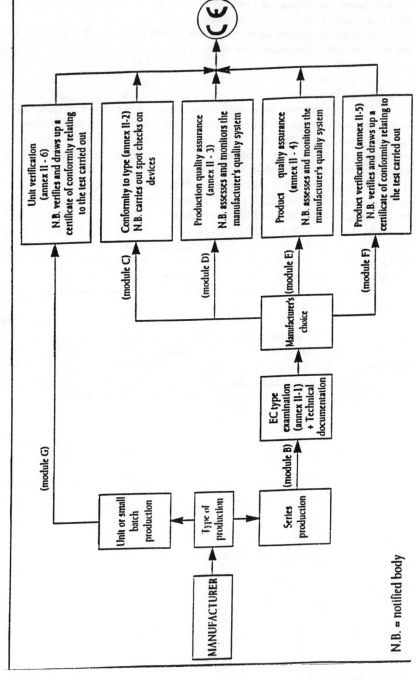

FLOWCHART FOR THE CONFORMITY ASSESSMENT PROCEDURES PROVIDED FOR IN DIRECTIVE 93/15/EEC ON EXPLOSIVES FOR CIVIL USE

N.B. = notified body

APPENDIX 1.

APPENDIX II

COMMISSION OF THE EUROPEAN COMMUNITIES
Director -General Industry.

STANDARDIZATION MANDATE TO CEN
FOR EXPLOSIVES FOR CIVIL USES.

1. BACKGROUND

Council Directive 93/15/CEE on the harmonization of the provisions governing the placing on the market and the supervision of explosives for civil use (O.J.N° L121 of May 1993) requires that explosives falling within the scope of the Directive must meet the safety requirements applicable to them before they can be placed on the market. The safety requirements are specified in Annex 1 of the Directive. Article 4 of the Directive lays down that explosives shall be presumed to comply with the essential safety requirements when they comply with the relevant national standards transposing the harmonized standards.

In July 1993 a programming mandate (M010) was given to CEN to draw up, in co-operation with CENELEC, a programme of standards giving safety requirements and test methods for explosives for civil uses, in line with the essential safety requirements of the Directive. CEN responded to this mandate on 7 February 1994 giving a list of standards which is annexed.

The purpose of this mandate is to invite CEN, in co-operation with CENELEC , to draw up such standards on the basis of the programme submitted in response to mandate M010.

The programming mandate emphasised the importance of the development of a common terminology. As well as overcoming an obstacle to harmonization, the elimination of differences in terminology and classification of explosives will contribute to the opening of barriers to trade and reduce the likelihood of unsafe practices. This aspect remains of crucial importance.

EUROPEAN COMMISSION

DIRECTORATE-GENERAL III
INDUSTRY
Directorate B: legislation and standardization; telematics networks
Unit III/B-2: standardization

Brussels, 11 April 1994
LP

Doc. 23/94 - EN

" STANDARDS AND TECHNICAL REGULATIONS " 83/189 CONSULTATIVE COMMITTEE

Title :   Draft standardization mandate to CEN for explosives for civil uses
             (M/055).

	Information		Discussion	X	Consultation

Rue de la Loi 200   -   B-1040 Brussels   -   Belgium
Telephone: Direct line: (+32) 2/295.66.00 - Secretariat: (+32) 2/296.19.06 - Standard: (+32) 2/299.11.11 - Fax: (+32) 2/296.89.98
Telex COMEU B 21877 - Telegraphic address COMEUR Brussels

List of standards, capable of giving presumption of conformity with the essential safety requirements of directive 93/15/EEC, as regards most civil explosives.

	Standards' title	Stage 32	Stage 40	Stage 49	Comments
	Standard A:				
	Explosives for civil uses				
1	Part I: Terms and definitions.	1994-12	1996-12	1998-06	
2	Part II: Classification.	1994-12	1996-12	1998-06	
3	Part III: Explosives for civil uses - Specification for labelling and other information to be provided by the manufacturer "Information requirement to include, where relevant, physical characteristics, chemical composition or purity; construction; dimensions; temperature of ignition; suitability at extreme temperatures; critical diameter; instructions for save handling and use; accessories; storage; shelf life; disposal".	1995-12	1997-12	1999-06	Test methods may have to be produced for a number of these properties e.g. Chemical composition, shelf life.

	Standards' title	Stage 32	Stage 40	Stage 49	Comments
	Standard B:				
	Explosives for civil uses - High explosives				
4	Part I: Requirements.	1996-12	1998-12	2000-06	It may be necessary to prepare separate specifications for each of the main types of product in this group.
5	Part II: Method(s) for the determination of thermal stability.	1995-12	1997-12	1999-06	
6	Part III: Method(s) for the determination of friction sensitivity.	1995-12	1997-12	1999-06	
7	Part IV: Method(s) for the determination of impact sensitivity.	1995-12	1997-12	1999-06	
8	Part V: Method(s) for the determination of initiation energy.	1995-12	1997-12	1999-06	
9	Part VI: Method(s) for the determination of propagation of detonation.	1995-12	1997-12	1999-06	
10	Part VII: Method(s) for the determination of performance characteristics. *Scope:* including, for example, density, velocity of detonation; power, etc.	1996-12	1998-12	2000-06	Final scope to be determined by the TC.
11	Part VIII: Methods (s) for the determination of critical diameter.	1995-12	1997-12	1999-06	

	Standards' title	Stage 32	Stage 40	Stage 49	Comments
12	Part IX: Method(s) for the determination of resistance to water.	1996-12	1998-12	2000-06	
13	Part X: Method(s) for the determination of safety at extreme temperatures.	1996-12	1998-12	2000-06	
14	Part XI: Method(s) for the determination of resistance to stress during loading of shotholes.	1996-12	1998-12	2000-06	
15	Part XII: Method(s) for the determination of resistance to hydrostatic pressure.	1995-12	1997-12	1999-06	
16	Part XIII: Explosives for use in underground works - Method(s) for the detection and measurement of toxic gases.	1996-12	1998-12	2000-06	
17	Part XIV: Explosives for use in coalmines or other underground works, where flammable gas may be a hazard - Method(s) for the assessment of suitability for use in flammable gases.	1996-12	1998-12	2000-06	
18	Part XV: Explosives for use in coalmines or other underground works, where combustible dust clouds may be a hazard - Method(s) for the assessment of suitability for use in combustible dust clouds.	1996-12	1998-12	2000-06	

	Standards' title	Stage 32	Stage 40	Stage 49	Comments
	Standard C: *Explosives for civil uses - Detonating cords, safety fuses and other fuses.*				
19	Part I: Requirements.	1996-12	1998-12	2000-06	It may be necessary to prepare separate specifications for each of the main types of product in this group.
20	Part II: Method(s) for the determination of thermal stability.	1995-12	1997-12	1999-06	
21	Part III: Method(s) for the determination of friction sensitivity.	1995-12	1997-12	1999-06	
22	Part IV: Method(s) for the determination of impact sensitivity.	1995-12	1997-12	1999-06	
23	Part V: Method(s) for the determination of resistance to accidental initiation.	1995-12	1997-12	1999-09	
24	Part VI: Method(s) for the determination of abrasion resistance.	1995-12	1997-12	1999-06	
25	Part VII: Method(s) for the determination of tensile strength.	1996-12	1998-12	2000-06	
26	Part VIII: Method(s) for the determination of velocity of reaction.	1995-12	1997-12	1999-06	
27	Part IX: Method(s) for the determination of initiation energy.	1995-12	1997-12	1999-06	
28	Part X: Method(s) for the determination of resistance to water.	1995-12	1997-12	1999-06	

	Standards' title	Stage 32	Stage 40	Stage 49	Comments
29	Part XI: Method(s) of for the determination of transmission of detonation.	1996-12	1998-12	2000-06	
30	Part XII: Method(s) for the determination of safety at extreme temperatures.	1996-12	1998-12	2000-06	
31	Part XIII: Method(s) for the assessment of suitability for use in hazardous environments.	1996-12	1998-12	2000-06	

	Standards' title	Stage 32	Stage 40	Stage 49	Comments
	Standard D: *Explosives for civil uses - Detonators and relays.*				
32	Part I: Requirements.	1996-12	1998-12	2000-06	It may be necessary to prepare separate specifications for each of the main types of product in this group.
33	Part II: Method(s) for the determination of thermal stability.	1995-12	1997-12	1999-06	
34	Part III: Method(s) for the determination of Impact sensitivity.	1995-12	1997-12	1999-06	
35	Part IV: Methods for the determination of resistance to accidental Initiation. *Scope:* Including, for example, drop test, electrostatic test, etc.	1995-12	1997-12	1999-09	Final scope to be determined by the TC.
36	Part V: Method(s) for the determination of initiating power output.	1995-12	1997-12	1999-06	
37	Part VI: Method(s) for the determination of delay accuracy. To include both pyrotechnic and electronic delays.	1995-12	1997-12	1999-06	
38	Part VII: Method(s) for the determination of resistance to water.	1995-12	1997-12	1999-06	
39	Part VIII: Method(s) for the determination of resistance to hydrostatic pressure.	1995-12	1997-12	1999-06	

	Standards' title	Stage 32	Stage 40	Stage 49	Comments
40	Part IX: Method(s) for the determination of mechanical integrity. *Scope:* Including, for example, tensile strength, abrasion resistance, etc.	1996-12	1998-12	2000-06	Final scope to be determined by the TC
41	Part X: Methods for the determination of safety under extreme temperatures.	1995-12	1997-12	1999-06	
42	Part XI: Detonators for use in coal mines or other underground works where flammable gas may be a hazard - Method(s) for the assessment of suitability for use in flammable gases.	1996-12	1998-12	2000-06	
43	Part XII: Electric and electronic detonators - Method(s) for the determination of electrical characteristics. *Scope:* Including, for example, no fire current, all-fire current, series firing current, minimum initiating energy, etc.	1995-12	1997-12	1999-06	Final scope to be determined by the TC.
44	Part XIII: Non-Electric detonators - Method(s) for the determination of shock tube characteristics. *Scope:* including, for example, reliability of ignition, velocity, etc.	1995-12	1997-12	1999-06	Final scope to be determined by the TC.
45	Part XIV: Definitions and requirements for devices and accessories for reliable and safe function of detonators and relays.	1996-12	1998-12	2000-06	

	Standards' title	Stage 32	Stage 40	Stage 49	Comments
	Standard E: _Propellants_				
46	Part I: Requirements.	1996-12	1998-12	2000-06	It may be necessary to prepare separate specifications for each of the main types of product in this group.
47	Part II: Method(s) for the determination of thermal stability.	1995-12	1997-12	1999-06	
48	Part III: Method(s) for the determination of friction sensitivity.	1995-12	1997-12	1999-06	
49	Part IV: Method(s) for the determination of impact sensitivity.	1995-12	1997-12	1999-09	
50	Part V: Method(s) for the determination of resistance to accidental initiation. _Scope_: Drop test, electrostatic testing, etc.	1995-12	1997-12	1999-09	
51	Part VI: Method(s) of test for deflagration to detonation transition.	1995-12	1997-12	1999-06	
52	Part VII: Method(s) for the determination of performance characteristics.	1996-12	1998-12	2000-06	
53	Part VIII: Method(s) for the determination of safety under extreme temperatures.	1996-12	1998-12	2000-06	

	Standards' title	Stage 32	Stage 40	Stage 49	Comments
54	Part IX: Solid rocket propellants - Method(s) for the determination of voids.	1996-12	1998-12	2000-06	
55	Part X: Solid rocket propellants - Method(s) for the determination of integrity of inhibitor coatings.	1996-12	1998-12	2000-06	

Accidents Involving Explosives and the Need for Safety Related Research

R. K. Wharton

HEALTH AND SAFETY LABORATORY, HARPUR HILL, BUXTON, DERBYSHIRE SK17 9JN, UK

1 INTRODUCTION

The Health and Safety Laboratory (HSL) is an in-house Agency of the Health and Safety Executive (HSE) and its role is to produce, acquire and interpret scientific information relating to occupational health and safety.

Currently our major customers are still the various HSE Inspectorates, although Agency status has enabled industry and other government departments to access and use the expertise of HSL on a contract basis. HSL's explosives work is carried out at the 550 acre Buxton site and is mainly for the Explosives Inspectorate, Mines Inspectorate and Quarries Inspectorate for whom we provide technical input to assist with their duties relating to HSE's mission of ensuring that the risks to people's health and safety from work activities are properly controlled.

The work at Buxton on explosives covers a wide range of products, e.g. commercial blasting agents and detonators, mining explosives and pyrotechnics. It also includes energetic substances (reactive chemicals such as sponge blowing agents and diazo salts) which are not intended to produce a primary explosive effect but can do so under certain circumstances.

In broad terms, HSL's activities in the explosives area can be classed as either reactive, planned or testing. Additionally, Explosives Section has responsibility for the Explosives Notified Body which, in response to the Civil Uses Directive, undertakes certification work connected with assessing whether explosives meet the essential safety requirements for placing on the market.

The title of this paper was chosen to encompass the main types of research that HSL is commissioned to do − reactive research, done to investigate particular incidents in the explosives industry, and planned research, which is undertaken to improve understanding of the hazards associated with specific work activities as a means of strengthening the basis from which advice and guidance can be given to industry.

Often research done as part of an incident investigation identifies the need for further work to quantify hazards, and currently about half of the research on explosives being undertaken at Buxton is of this incident-driven type.

Examples are selected from all three categories to illustrate the way that targeted safety related research has improved our understanding of the hazards associated with certain industrial processes involving explosives.

2 REACTIVE RESEARCH

Historically, a significant proportion of HSL's work on explosives has been devoted to the investigation of accidents and incidents. In the thirteen year period from 1980 - 1993 the Section investigated twenty-eight separate incidents relating specifically to mining use and seventy-six incidents in the non-mining sector. The paper by Thomson gives more details of certain of these investigations[1].

2.1 The Explosion at Peterborough

Included in the list is the technical investigation of the explosion at Peterborough in March 1989 that involved the detonation of a mixed load of about 800 kg of blasting explosives and detonators.

This accident occurred when the driver of a laden explosives vehicle lost his way during routine deliveries and entered an industrial estate to ask directions. As he travelled over a speed control ramp going into the estate he heard a dull thud and, in his wing mirrors, saw a tongue of flame emitted from one side of the rear of the vehicle. Subsequent examination showed that the rear roller shutter door had become dislodged and that a fire was developing in the cargo. He then raised the alarm and surrounding buildings were evacuated: the full cargo detonated some twelve minutes later causing substantial damage to surrounding buildings, injuring over eighty people and resulting in one fatality.

The in-depth investigation undertaken by the Explosives Inspectorate examined blast damage and the safety management system of the company, and identified that one item in the load was being transported in unapproved and unauthorised packaging. HSL was subsequently asked to determine the mechanism for initiation. After detailed examinations of the stability and sensitiveness of the various items comprising the load, attention focused on the cerium fusehead combs that were being transported in an unapproved manner.

It was demonstrated that, with the combs packaged as they were in the original cargo, it was possible to detach pieces of composition during the vibration experienced in simulated road transport. This composition was quite sensitive to mechanical forces (impact, friction) and was shown to be greatly sensitised by the incorporation of small amounts of rust such as were present in the inner steel transport boxes.

Work done to measure the mechanical energy transferred to packages when a loaded vehicle travelled over a speed control ramp indicated that for the rust-sensitised material it would have been possible to produce energy inputs equivalent to those that had resulted in initiation in small scale laboratory tests.

Subsequent simulation trials, with a van containing fusehead combs packaged as they were prior to the incident, indicated that a significant fireball was produced on ignition of the pyrotechnic. With the box at the location it occupied in the original load, the forces

Figure 1 *Simulation trial illustrating the result of initiating a box of cerium fusehead combs in a vehicle*

generated by fireball production were sufficient to dislodge the rear roller shutter door allowing the fireball to vent as a jet from the nearest side of the vehicle, Figure 1.

This replicated witness statement observations, as did the subsequent fire development, since the flames spread to empty wooden boxes which were stored next to the fusehead combs, and then on to boxes of detonators. The twelve minute delay to detonation is accounted for by the relative position of items of the cargo in relation to the cerium fusehead combs. The functioning of a number of the detonators produced the 'pops' which were reported by witnesses. Eventually the containment of the detonator boxes was breached and tests showed that some live units would have been strewn on to the remaining cargo which comprised heated blasting explosives. Mass detonation of the load followed shortly afterwards.

In this investigation research was able to demonstrate a plausible initial initiation mechanism, provide simulation trials to demonstrate large scale effects in keeping with witness observations, and account for the observed time delay to detonation. A fuller account of this work can be found elsewhere[2].

2.2 Ignition of Titanium / Blackpowder Pyrotechnic Compositions

The pyrotechnic mixture of titanium metal and blackpowder is used in certain display fireworks known as gerbs. The manufacturing process for these relatively large (typically 3 kg) fireworks involves consolidating increments of composition by pressing, and two separate incidents had occurred (the second resulting in a fatality) in which there were accidental ignitions.

As part of our investigation of the causes of these accidents we examined the sensitiveness of the titanium/blackpowder mix to mechanical forces. Mixtures containing up to 30% metal are commonly used in industry but we extended the range of relative proportions of ingredients in order to fully examine the nature of the dependence of reactivity on composition.

Initial results indicated that, over the size range examined, the particle size of the titanium exerted no effect on mechanical sensitiveness.

However, both impact and friction sensitiveness were shown to be dependent on the quantity of titanium in the composition, and for friction the dependence took the form illustrated in Figure 2. These results indicate that compositions containing 15 - 30% titanium are the most likely to respond to frictional forces. Since this is the titanium content generally used in industry, the results offer an explanation for the occurrence of ignitions during mechanical processing. Of more interest however is the observation that re-formulation to mixtures containing ca. 8% titanium gives the potential for considerable safety gains during processing.

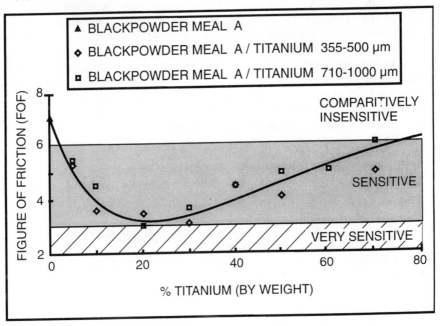

Figure 2 *The effect of titanium metal content on the friction sensitiveness of titanium / blackpowder mixtures*

Industry has now adopted such an approach with only a marginal loss in performance of the fireworks and the added gain of a reduction in the quantity of relatively expensive titanium needed for manufacture.

The increase in friction sensitiveness found over the range of titanium 0 - 25% is in keeping with the known sensitisation of explosives by the incorporation of hard gritty particles. At titanium levels above ca. 30% the increased quantity of metal acts as a heat sink, thus reducing reactivity as reflected in the form of the dependence in Figure 2.

Full details of this work have been published elsewhere[3,4].

3 INCIDENT DRIVEN RESEARCH

3.1 Firework Mortars

An incident occurred during the firework display to mark the end of the Glasgow Garden Festival in 1988[5]. The main charge of a 200 mm (8") firework display shell exploded before the lifting charge i.e. while the shell was still in the metal launch tube. The mortar tube was shattered by the explosion and shrapnel was thrown over a considerable distance causing injury to members of the public and to the display operator who lost his leg.

Examination of the internal fusing arrangements of an equivalent firework at the laboratory was able to show that the bursting charge and lifting charge had functioned simultaneously since the flight delay fuse had failed. The incident raised an awareness of the potential of such powerful fireworks to cause injury to firework operators (an at work activity) and to members of the public, and a research programme was therefore initiated in order to both quantify the hazards and to examine methods of mitigation. The need for such work has been reinforced by subsequent similar accidents in Japan[6] and the USA[7].

The project has investigated the behaviour of spirally wound and straight tube steel mortars, and the role of the weld[8] in straight tubes. The number and energies of fragments produced on misfire from different sized tubes have also been quantified.

Technical information of this type could be used when safety distances are examined in future revisions of guidance relating to firework displays.

In relation to mitigation, the effectiveness of rubber tyre collars, sandbags and sand barrels have been evaluated and ranked. The research may be extended to examine the different performance of alternatives to steel for mortar tube construction e.g. certain modern plastics (such as high density polyethylene) can withstand the temperatures and pressures generated by a normal shell launch but will bulge and split to release the gases formed on misfire rather than fragment to produce dangerous high velocity projectiles.

There could also be a role for industry in a collaborative project, especially in connection with the supply of materials for test. For example, one as yet unresolved issue relates to the degree of change in performance of metal mortar tubes with repeated field use.

3.2 Fires Involving Pyrotechnics

A fire occurred in 1989 in a fireworks production facility during cleaning operations. Individual boxes of composition were being moved across a workbench when the contents of one ignited producing a large fireball. Both of the workers who were present in the small building at the time were severely burned and one subsequently died from her injuries.

The cause of this fire was investigated at Buxton and ascribed to the trapping of a piece of mechanically sensitive star composition under one of the boxes. This was subsequently initiated by friction as the box was moved, causing the contents to ignite.

In order to obtain a relative ranking of the threats posed by the fireballs produced on the ignition of the various quantities of different compositions present in the process building, a number of trials were done in which fireball size and duration were measured.

This initial work[5] indicated that, in common with certain other explosives properties (notably overpressure), the diameter and duration of the fireball were both approximately proportional to $M^{1/3}$, where M is the net explosive content of the composition.

Subsequently a research project was commissioned with the aim of extending both the range of pyrotechnic compositions examined and the values of M. Measurements were taken not only of fireball diameter and time-scale but also of the thermal emissive power

Figure 3 *Fireball produced on the ignition of 25 kg of gunpowder (inner marker poles are 3 m high)*

of the flames. The study produced a substantial quantity of new data[9] and showed that the fire hazard posed by certain pyrotechnics can be substantial. For example, Figure 3 illustrates the fireball produced on ignition of 25 kg of gunpowder under conditions of self confinement.

These data have recently been used to evaluate thermal dose and hence the distances to various degrees of burn injury[10]. Information of this type relating to quantification of the potential hazards from the typical quantities of pyrotechnics that could be encountered in industrial manufacturing, handling and processing situations is the first step towards providing effective safeguards against the associated risks.

Although engineering controls are the primary method of protecting workers from industrial hazards, as a last resort personal protective equipment is used. Current practice in relation to fire protective clothing in the pyrotechnics manufacturing industry was recently reviewed[11] and in November 1995 the Explosives Industry Group of the CBI produced a guide to the selection, maintenance and use of fire protective clothing in the pyrotechnics and propellants industries[12]. Both studies identified the need for a realistic means of assessing fire protective clothing for use in the explosives industry.

The study of pyrotechnic fireballs has provided quantification of the thermal threat to which workers could be exposed: the work that remained was to use this information to

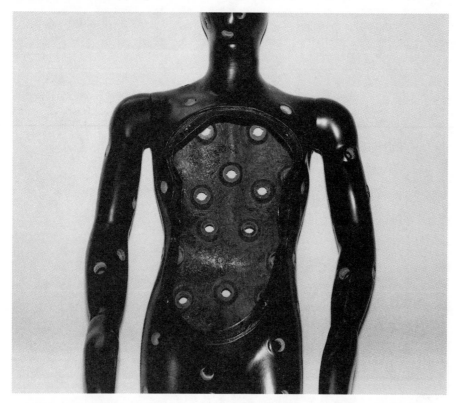

Figure 4 *Manikin for performance testing of fire protective clothing for explosives workers (with chest cover plate removed) illustrating positions for heat sensors*

select a range of source terms to enable fire protective clothing to be evaluated in a realistic manner.

The Explosives Inspectorate has recently commissioned an extramural research programme with the British Textile Technology Group to develop a full torso manikin test to be used for evaluating the performance of complete fire protective clothing systems, Figure 4. The contract, which is being managed by HSL, involves producing a fully instrumented manikin that will react to fire situations by turning and retracting. Response times and escape velocities used in HSE's major hazards assessment work have been used as the design criteria.

This work area links a number of projects together and provides a good example of how HSE co-ordinates research to meet a broad need.

From the initial incident investigation through to the evaluation of clothing systems for the explosives industry under realistic conditions, the research work is strengthening the technical background from which additional advice can be given and guidelines can be formulated.

3.3 Handling of Initiating Devices

Certain detonators and primers are handled routinely in industry while others require the use of remote manipulation by robots. Unfortunately, there have been a number of incidents in the UK involving accidental ignition of initiating devices during manual

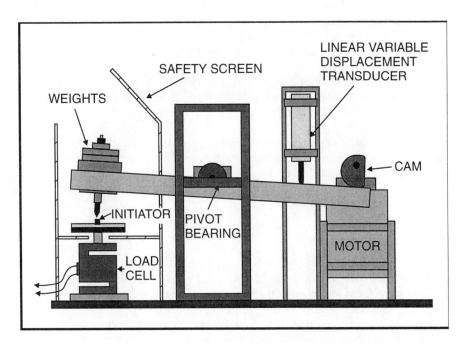

Figure 5 *Apparatus for evaluating the response of initiating devices to mechanical forces*

handling and a research study was therefore initiated in order to study the forces required to function these devices.

The extramural study involved the design of an apparatus for measuring the forces exerted by the human hand, its use in a typical industrial environment, and the construction of a laboratory test to replicate the measured forces[13, 14]. Use of the latter apparatus provides a means of identifying those initiating devices that present the greatest hazard during normal manual handling and also, by increasing the applied force above that exerted by the human hand until ignition occurs, it is possible to obtain an indication of the relative margin of safety involved in manually handling a range of devices.

Figure 5 illustrates the apparatus used to replicate industrially measured forces and to establish the minimum forces needed to function the devices.

4 PLANNED RESEARCH

4.1 Vehicle Air Bags

Current vehicle air bags are mainly devices driven by the functioning of an explosives charge (sodium azide or nitro-glycerine / nitro-cellulose propellant).

Whereas the way the devices function in a car to protect the driver, and also increasingly the front seat passenger, is a consumer safety matter, there are areas concerned with the introduction of this new technology that come within the remit of HSE i.e. the work-related issues of transport, storage, classification, handling and disposal.

In order to evaluate the potential hazards to workers and provide a technical background from which guidance could be written, HSL was asked to undertake a number of trials.

The work has involved assessing the hazard to line-side car workers posed by the accidental initiation of an unrestrained air bag module, the effectiveness of air bag storage cabinets in the event of a fire, and the hazards posed by a fire in a vehicle transporting air bags[15].

Future work will (i) extend the range of packages of air bags to be examined in order to determine how they respond in an engulfing fire, and (ii) examine the hazards posed to emergency services personnel by the presence of an uninitiated explosive device i.e. when they are cutting survivors from a crashed car in which the air bag has not functioned (e.g. an accident involving side impact).

5 CONCLUSIONS

The above brief overview of some of HSL's recent and current research relating to explosives provides an indication of the need for and benefits to be gained from technical safety-related studies.

Both reactive and planned research have a part to play in supporting the activities of HSE's various Inspectorates and often the work done in the investigation of accidents can identify the need for a more fundamental applied research study.

Since it is HSE's policy to publish the results of safety related research, many of the outcomes of individual studies can be found in appropriate scientific journals and in conference proceedings.

6 REFERENCES

1. B. J. Thomson, *Expl. Eng.*, 1994, **June,** 4.
2. R. K. Wharton and R. J. Rapley, *Prop., Expl., Pyro.*, 1992, **17**, 139.
3. R. K. Wharton, *Proc. 1st Intl. Symp. on Fireworks, Montreal, Canada*, 1992, 339.
4. R. K. Wharton, R. J. Rapley and J. A. Harding, *Prop., Expl., Pyro.*, 1993, **18**, 25.
5. R. K. Wharton, *Proc. 16th IPS, Jönköping, Sweden*, 1991, 514.
6. B. J. Thomson, Fireworks Mortar Shell Accidents in Japan on 2 August 1989 & 5 August 1989, RLSD Report IR/L/SM/90/03.
7. *Fireworks Business*, 1994, **126**, 3.
8. S. G. Myatt, *J. Pyro.*, 1995, **1**, 6.
9. R. K. Wharton, J. A. Harding, A.J. Barratt and R. Merrifield, *Proc. 21st IPS, Moscow, Russia,* 1995, 916.
10. R. Merrifield, R. K. Wharton and S. A. Formby, US Dept. of Defence, Explosives Safety Board, *27th Explosives Safety Seminar*, 20 - 22 August, Sahara Hotel, Las Vegas, 1996.
11. N. I. Sorensen and R. K. Wharton, *Proc. 16th IPS, Jönköping, Sweden,* 1991, 747.
12. "Fire Protective Clothing", CBI Explosives Industry Group, Confederation of British Industry, November 1995 (ISBN 0 85201 513 5).
13. S. C. Campbell, *Expl. Eng.*, 1996, **March,** 12.
14. S. C. Campbell, P.F. Nolan, R. K. Wharton and A. W. Train, *Proc. 21st IPS, Moscow, Russia*, 1995, 109.
15. A. E. Jeffcock, *Proc. Explo '96, Leeds*, 1996, 19.

Session 3. Security and High Energy Explosives

Modern Explosives and Nitration Techniques

N. C. Paul

DEFENCE RESEARCH AGENCY, FORT HALSTEAD, SEVENOAKS, KENT TN14 7BP, UK

1 INTRODUCTION

The number of explosive materials in military use at the time of Alfred Nobel's death in 1896 was limited and munitions were mainly based on nitroaromatic and nitrate ester compounds, such as; trinitrotoluene (1), nitroglycerine (2) and nitrocellulose (3) and though some of these materials are still in use today they have been largely superseded by more powerful explosives.

In this context, the term explosives will include 'low explosives' or propellants for gun and rocket applications. Over the last 100 years there has been considerable development of military explosives with the introduction of many new explosive compounds and the most important of these are shown in Table 1. In addition to the introduction of new energetic compounds, modern formulations rely heavily on the inclusion of other materials to enhance the physical properties of compositions to meet a range of specific uses and service conditions. For example, modern compositions include binders to give improved physical integrity and greater safety in handling and in order to maintain high performance these materials are often energetic materials in their own right. A current example is the development of energetic binders which contribute a significant portion to the overall energy of the composition. For military use the prime driving force in the development of explosives is higher and higher performance but with a drive towards safer, less sensitive materials. In addition to these two principal drivers, environmental concerns are now becoming increasingly important and this latter aspect is being considered throughout the life cycle of an energetic material; from production, through

Table 1 *Major Explosives Development*

Explosive	First Reported	In Service
	(approximate dates)	
TNT	1863	~1900
TETRYL	1877	1906
symTNB	1887	1910
Picric Acid	~1900	1910
PETN	~1900	~1930
RDX	1900	~1935
HMX	~1930	~1970

Table 2 *Waste Acid Generated in the Manufacture of Explosives*

Compound	Nitration System	kg spent acid per kg product
TNT (overall)	Sulphuric/Nitric	1.5
RDX	Nitric acid	1.4
RDX	Nitric/Acetic	1.6
HMX	Nitric/Acetic	1.8
NG	Sulphuric/Nitric	1.4
NC	Sulphuric/Nitric	20-30

formulation, storage and potential use and extends to the disposal of munitions reaching the end of service life. No discussion on the subject of explosives would be complete without reference to nitration processes which are integral to their production. Nitration using traditional nitrating agents is well documented but this paper will also present recent work on the application of dinitrogen pentoxide chemistry to explosives production.

2 HIGH PERFORMANCE EXPLOSIVES

Chemical explosives obtain their energy as the result of chemical reactions which transform an initially metastable molecule into one or more molecules of greater stability. The reactions are very fast and this distinguishes explosive decomposition from normal combustion processes. The measurement of performance is dependent on the application and different parameters will apply to high explosives, gun propellants and rocket propellants. Thus, for a high explosive the major effect required, blast or metal moving, is best measured by the Detonation Pressure (P_{CJ}) and Velocity of Detonation (VoD); whilst for gun and rocket propellants a measure of propelling power is provided by the Specific Impulse (I_{sp}). A further indication of performance can be gained from the calculation of an oxygen content sufficient to convert all the carbon to carbon dioxide and the hydrogen to water. This is especially important for propellant compositions. Major improvements in explosive performance have been achieved through the introduction of nitramine compounds; the two most notable materials being RDX (4) and HMX (5) developed in the early 1940 s - 50 s.

(4)   (5)

HMX has been used in explosive compositions since the 1950s and is still currently the most powerful explosive in service but a relatively recent addition to this class of explosive materials is HNIW, hexanitrohexazaisowurtzitane (6).

(6)

This explosive has a higher density and a greater explosive power than HMX and is now being evaluated for a number of explosive applications. It has been calculated that with this compound we are nearing the theoretical maximum explosive power that can be attained with conventional C,H,N,O compounds. Attempts are being made to produce stable molecules with high ring strain that may increase the power output and two target molecules are; octanitrocubane (7) and octa-azacubane (8).

Octa-azacubane, if it can be made, is calculated to be the 'ultimate explosive'.

3 INSENSITIVE EXPLOSIVES

Inadvertent initiation of munitions can occur through accidental shock, fragment impact or as the result of a fuel fire and, following a number of major incidents involving the accidental detonation of munition stores, the attention of munition designers turned to the problem of safety, culminating in a national requirement for insensitive munitions.

Reduction in overall sensitivity can be brought about by actual weapon design to resist, for example, the effect of mechanical shock In addition, there has been considerable research on the synthesis of less sensitive compounds without compromising the high performance requirements. Two materials of note are; HNS, hexanitrostilbene (9) and TATB, triaminotrinitrobenzene (10).

HNS is an high melting explosive with a high decomposition temperature and has been used in deep mining and space applications. TATB is a very insensitive explosive and this is thought to be due to stabilisation of the molecule through the formation of internal hydrogen bonds between the amino and nitro groups. This is an area of on-going research, especially in the synthesis of high nitrogen heterocyclic compounds.

Many explosives can also be rendered less sensitive by the inclusion of desensitising agents in the explosive charge. An early example of this was the addition of beeswax to TNT in warhead compositions. It is more common nowadays to incorporate the explosive in a rubbery matrix which can absorb the mechanical shock and this has been demonstrated by bullet impact tests on simulated rocket motors as shown in Figure 1.

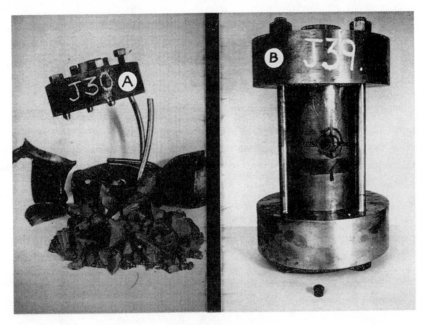

Figure 1 *Bullet Impact Test on Simulated Rocket Motor*
A) Brittle Propellant   B) Rubbery Propellant

The test on a brittle composition resulted in a detonation with destruction of the metal casing whereas, in the case of a 'rubbery' propellant there was ignition and burning but no detonation. A binder material, currently in use, is an hydroxyterminated *poly* - butadiene rubber but, being inert, the binder dilutes the overall energy of the composition leading to a lower performance. This has been addressed by the synthesis of energetic binders based on oxetanes, typified by *poly* -NIMMO(11):

$$HO-CH_2-\underset{\underset{CH_3}{|}}{\overset{\overset{CH_2ONO_2}{|}}{C}}-O\left[CH_2-\underset{\underset{CH_3}{|}}{\overset{\overset{CH_2ONO_2}{|}}{C}}-O\right]_n CH_2-\underset{\underset{CH_3}{|}}{\overset{\overset{CH_2ONO_2}{|}}{C}}-OH$$

(11) *poly* -Nitratomethyl Methyl Oxetane (*poly* -NIMMO)

and those based on oxiranes, e.g. *poly* -GLYN (12):

$$
\begin{array}{ccc}
ONO_2 & \left[\ ONO_2\ \right] & ONO_2 \\
| & | & | \\
CH_2 & CH_2 & CH_2 \\
| & | & | \\
HO\text{-}CH_2\text{-}CH\text{-}O\text{---}CH_2\text{-}CH\text{-}O\text{---}CH_2\text{-}CH\text{-}OH
\end{array}
$$

(12) *poly* -Glycidyl Nitrate (*poly* -GLYN)

Compositions are produced by incorporating the explosive filler in the liquid prepolymer which is then crosslinked to an elastomeric polyurethane rubber with isocyanates. A number of explosive and propellant formulations have been produced with these new materials which show great promise in terms of both increased energy and lower vulnerability.

4 ENVIRONMENTAL FACTORS

Since the crosslinking of polyurethane binders is irreversible compositions cannot be reprocessed, and any defects in manufacture cannot be rectified, leading to considerable wastage and problems in the disposal of end of service munitions.

A new generation of energetic thermoplastic binders, which can be reprocessed, is in development. These have good low temperature rubbery properties but can also be reversibly melted for formulation, re-formulation and ease of demilitarisation. This latter aspect increases the possibility of recovering and recycling explosive components which are costly to produce and represent a significant investment.

The major level of environmental pollution however, occurs at the start of the explosives life cycle during manufacture of explosives. Since the majority of energetic compounds in use contain nitro-substituents (C-NO$_2$ nitrocompounds, O-NO$_2$ nitrate esters, N-NO$_2$ nitramines), nitration processes play a vital role in their production. Nitrations are generally notoriously polluting processes, generating large quantities of waste acids which are costly to recover and treat and this is indicated in Table 2 which shows the level of waste acid generated in the manufacture of a number of explosives. Cleaner processes and reduced waste streams are necessary if future production is to meet the increasing environmental legislation.

Work at the Defence Research Agency, started in the early 1980s, on the production and use of dinitrogen pentoxide has shown that there is a significant environmental benefit since dinitrogen pentoxide possesses some unique advantages over conventional nitrating agents in the cleaner production of energetic materials with reduced waste acid streams and cleaner processes.

5 NITRATIONS WITH DINITROGEN PENTOXIDE

This paper will now describe some of the advances in nitration techniques and, especially, recent work on dinitrogen pentoxide (N_2O_5) chemistry. Although dinitrogen pentoxide was first discovered in the late 1890s it is only recently that satisfactory production processes for manufacture have been established. Dinitrogen pentoxide is a white crystalline solid, soluble in nitric acid and inert chlorinated solvents. It is metastable, slowly decomposing at room temperature to dinitrogen tetroxide and oxygen, and, being the anhydride of nitric acid, reacts   readily with water to form two molecules of nitric acid. It is these two properties which have created past difficulties in obtaining the pure material in sufficient quantities for commercial use. Two methods have now been developed for large scale production; ozonation of dinitrogen tetroxide and an electrolytic route.

5.1 The Ozonation Process

In this process a stream of ozone in air or oxygen is contacted with dinitrogen tetroxide gas such that the ozone is in slight stoichiometric excess. Reaction is instantaneous and the dinitrogen pentoxide produced is trapped out at low temperature and can be stored at temperatures of -50 to -60°C for long periods without decomposition. The solid product can be dissolved in a solvent of choice for reaction. A variant of this process, suitable for larger quantities, has been developed where the reaction takes place within a suitable solvent.

5.2 The Electrolytic Process

Electrolysis of dinitrogen tetroxide dissolved in nitric acid in a divided cell results in the oxidation at the anode to form dinitrogen pentoxide and reduction of nitric acid at the cathode to form dinitrogen tetroxide. The overall cell process is the dehydration of nitric acid (Scheme 1).

$$2HNO_3 + N_2O_4 = 2N_2O_5 + 2H^+ + 2e^- \quad \text{ANODE}$$
$$2HNO_3 + 2H^+ + 2e^- = N_2O_4 + 2H_2O \quad \text{CATHODE}$$

$$2HNO_3 = N_2O_5 + 2H_2O \quad \text{Overall}$$

Scheme 1

The overall efficiency of this process can be maximised by collecting and recycling the dinitrogen tetroxide produced in the cathode compartment. The electrolysis equipment is based on a frame and plate configuration and is very amenable to scale-up for very large production. Dinitrogen pentoxide from this route is produced as a solution in nitric acid but, under current development is a process whereby the dinitrogen pentoxide can be extracted, in pure form, and subsequently dissolved in an organic or other solvent.

5.3 Nitrations

Dinitrogen pentoxide exhibits a dichotomy of behaviour according to the medium in which it is employed. In 100% nitric acid a high nitronium ion concentration is present on account of the high degree of dissociation of N_2O_5 arising from the polarity of the solvent whereas, in organic solvents, typically chlorinated hydrocarbons, the N_2O_5 is essentially undissociated.

5.3.1 Dinitrogen Pentoxide in Nitric Acid. This medium provides a potent, unselective nitration system akin to mixed acid in nitrating power. Nitration of aromatic compounds, typified by the examples below (Scheme 2), is essentially quantitative and the products are applicable to explosive applications.

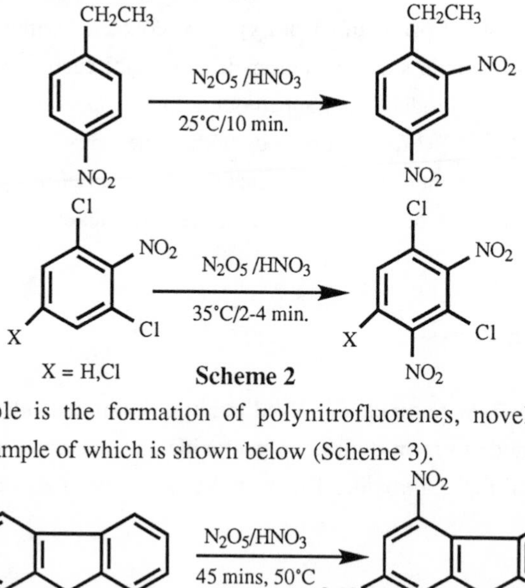

X = H,Cl   **Scheme 2**

A further example is the formation of polynitrofluorenes, novel thermally-stable explosives, an example of which is shown below (Scheme 3).

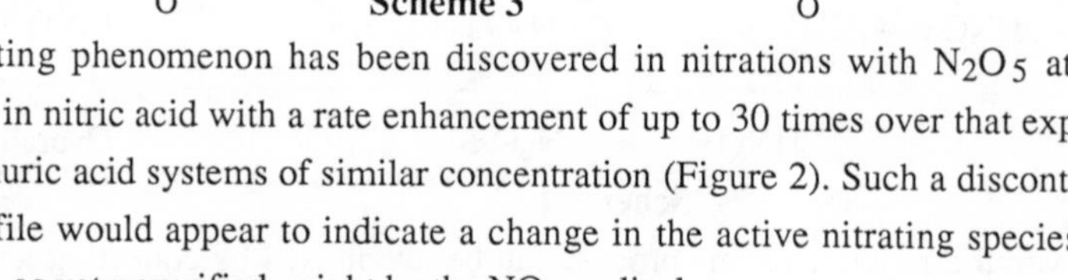

Scheme 3

An interesting phenomenon has been discovered in nitrations with N_2O_5 at high concentrations in nitric acid with a rate enhancement of up to 30 times over that expected for nitric-sulphuric acid systems of similar concentration (Figure 2). Such a discontinuity in the rate profile would appear to indicate a change in the active nitrating species; one such candidate, as yet unverified, might be the $NO_3\cdot$ radical.

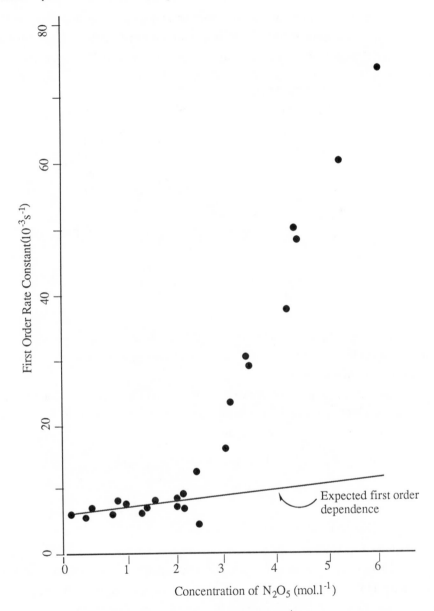

Figure 2 *Rate profile for the nitration of phenyltrimethylammonium perchlorate by N_2O_5 / HNO_3*

Nitrolysis reactions to form nitramines are important reactions in explosives synthesis and N_2O_5 offers considerable improvements over existing methodology, largely due to the instability of these products in media containing sulphuric acid. This route has been applied to the production of pure HMX via DADN (Scheme 4).

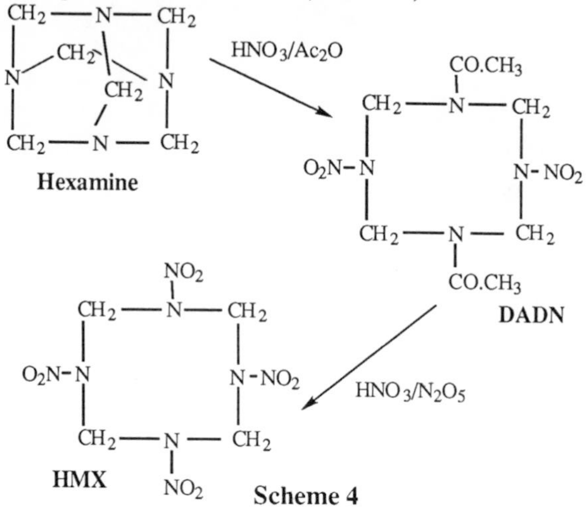

Scheme 4

5.3.2 Dinitrogen Pentoxide in Organic Solvents. Reactions in organic solvents embrace two distinct but related areas of chemical synthesis, namely, ring-cleavage reactions and selective nitrations.

Ring cleavage with dinitrogen pentoxide now represent a general class of nitration reaction enabling the synthesis of a large number of energetic compounds (Scheme 5) which have applications in explosives technology.

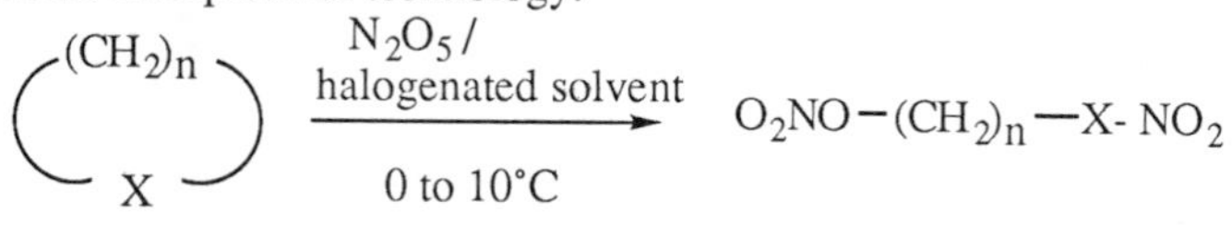

$$n = 2 \text{ or } 3$$
$$X = O \text{ or } NR \, (R \neq H) \qquad \textbf{Scheme 5}$$

The substrates are strained-ring heterocycles with three- or four-membered rings containing either oxygen or nitrogen heteroatoms. It should be noted that these reactions possess a common feature in that there is simultaneous addition of two energetic groups; nitramine-nitrates in the case of nitrogen heterocycles and di-nitrates in the case of oxygen heterocycles with complete utilisation of the nitrating agent. The reactions are generally high yielding with reduced by-products compared with conventional nitration reactions.

Owing to the mild nitrating power of dinitrogen pentoxide in an organic solvent medium it is possible to allow a nitration to proceed only partially where two or more functional groups of widely differing lability are present in the same molecule.

Such an approach would not be possible with conventional strongly acid media, e.g. nitric- sulphuric acid mixtures, whilst other known mild reagents, such as acetyl nitrate or tetranitromethane, are impractical on a large scale on account of excessive hazard and expense. The order of reactivity for various nitratable moieties , based on experimental observations is: olefinic double bond< aromatic substitution < oxetanes and azetidines < oxiranes and aziridines < hydroxyl. A further feature of dinitrogen pentoxide in organic solvent media is that reactive polymer chains can be nitrated without chain scission of the polymer backbone. An example of these features is seen in the partial nitration of a polybutadiene in which a portion of the double bonds had been converted to oxirane groups. If only sufficient N_2O_5 to react with the oxirane groups is used ring cleavage of these groups is the only reaction observed (Scheme 6).

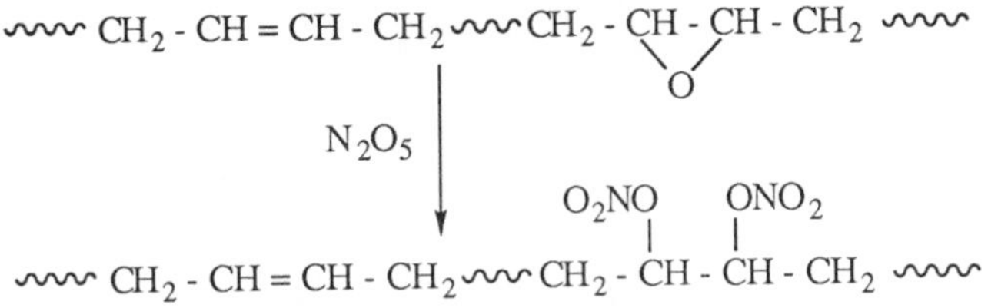

$$\sim\!\!\sim\!\!\sim CH_2 - CH = CH - CH_2 \sim\!\!\sim\!\!\sim CH_2 - \underset{\diagdown\!\!\diagup}{CH} - \underset{O}{CH} - CH_2 \sim\!\!\sim\!\!\sim$$

$$\Big\downarrow N_2O_5$$

$$\sim\!\!\sim\!\!\sim CH_2 - CH = CH - CH_2 \sim\!\!\sim\!\!\sim CH_2 - \underset{O_2NO}{CH} - \underset{ONO_2}{CH} - CH_2 \sim\!\!\sim\!\!\sim$$

Scheme 6

This material, containing vicinal dinitrate groupings, is a potential energetic binder for explosive compositions. The energy content can be controlled by varying the degree of epoxidation of the polybutadiene precursor up to a maximum of about 30% of the original double bonds. Attempts to increase the degree of epoxidation leads to the formation of ester and carbonyl by-products which are undesirable.

A second approach to the synthesis of energetic binders also makes use of the selectivity of N_2O_5 / organic solvent properties in that energetic monomers for subsequent polymerisation can be produced from hydroxy substituted oxiranes and oxetanes. The two energetic binders mentioned previously; *poly*-NIMMO (11) and *poly*-GLYN (12) are produced by polymerisation of the corresponding NIMMO and GLYN monomers which, in turn are prepared by selective nitration of the hydroxyl functions of their respective precursors (13, 14). Nitration takes place at low temperature and conversion of the hydroxyl group is complete within a few seconds. If an excess of N_2O_5 is used and the reaction time extended further nitration through ring-opening can take place (Scheme 7).

Purity of the monomers is of paramount importance for the polymerisation step and it is essential that reactions are carefully controlled to minimise the ring-opening nitration. A further complication is that in the nitration of the hydroxyl group a molecule of nitric acid is also produced and this may also cause ring-opening with the formation of a nitrate and a further hydroxyl group (Scheme 8).

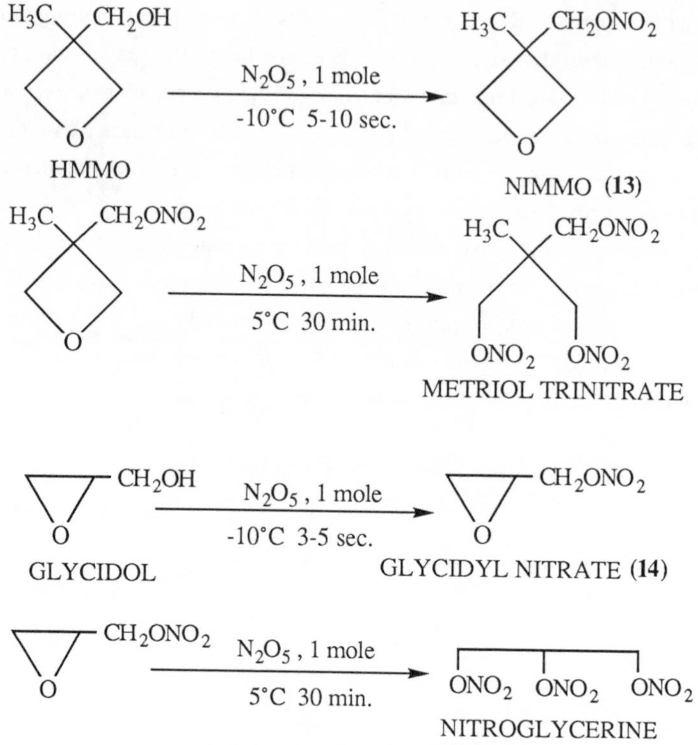

H₃C, CH₂OH

$$\xrightarrow[\text{-10°C 5-10 sec.}]{\text{N}_2\text{O}_5\text{ , 1 mole}}$$

H₃C, CH₂ONO₂

HMMO

NIMMO (13)

H₃C, CH₂ONO₂

$$\xrightarrow[\text{5°C 30 min.}]{\text{N}_2\text{O}_5\text{ , 1 mole}}$$

H₃C, CH₂ONO₂

ONO₂   ONO₂

METRIOL TRINITRATE

CH₂OH

$$\xrightarrow[\text{-10°C 3-5 sec.}]{\text{N}_2\text{O}_5\text{ , 1 mole}}$$

CH₂ONO₂

GLYCIDOL

GLYCIDYL NITRATE (14)

CH₂ONO₂

$$\xrightarrow[\text{5°C 30 min.}]{\text{N}_2\text{O}_5\text{ , 1 mole}}$$

ONO₂   ONO₂   ONO₂

NITROGLYCERINE

Scheme 7

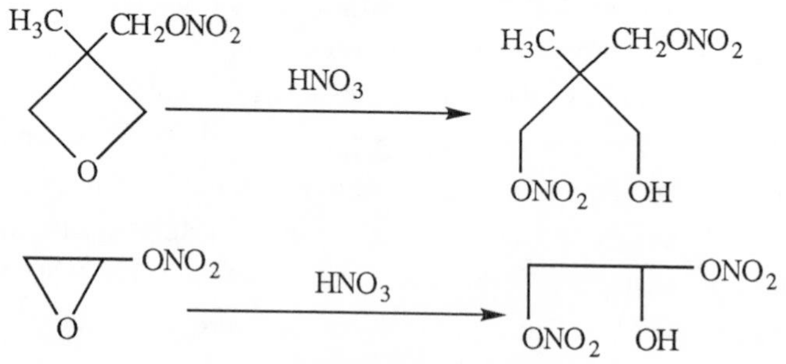

H₃C, CH₂ONO₂

$$\xrightarrow{\text{HNO}_3}$$

H₃C, CH₂ONO₂

ONO₂   OH

ONO₂

$$\xrightarrow{\text{HNO}_3}$$

ONO₂

ONO₂   OH

Scheme 8

Pure materials can be produced on a batch laboratory scale but scale-up to provide useful quantities is difficult. Nitration of the hydroxyl group is highly exothermic resulting in longer addition times necessary to contain the exotherm. This has been overcome by nitration in a continuous flow reactor followed by immediate quenching with aqueous carbonate solution and large quantities of exceptionally pure monomers can now be produced in near quantitative yields.

6 SUMMARY

During the 20th century, explosives have developed from simple compounds into highly complex formulations to meet changing requirements and a variety of new applications. The main drive has always been to higher performance compositions but safety has become increasingly important, often leading to compromises. As we progress towards the next century a third driver, now coming very much to the fore, is the environmental factors and which must cover the whole life cycle of a munition. Environmental legislation, driven by public awareness, is becoming stricter and many older manufacturing processes may not be permitted in the future. The development of dinitrogen pentoxide nitration technology and its application to the production of explosive ingredients marks a significant step towards the cleaner manufacturing processes which will need to be developed. This technology, is by no means limited to the production of explosive ingredients but has implications for the whole chemical industry. In addition to the environmental factors there are a number of materials that can be produced using this reagent, that would not otherwise be available.

Dinitrogen pentoxide nitrations are versatile and have much to offer in the area of energetic materials; the same technology should also offer advantages in other synthesis and chemical process areas.

© British Crown Copyright 1996/DERA
Reproduced with the permission of the Controller of
Her Britannic Majesty's Stationery Office

Explosives Detection by Ion Mobility Spectrometry

Rod Wilson and Alan Brittain

GRASEBY DYNAMICS LTD, PARK AVENUE, BUSHEY, WATFORD, HERTFORDSHIRE
WD2 2BW, UK

1. Summary.

Having the characteristics of high sensitivity, fast response times and hand portability, Ion Mobility Spectrometry (IMS) has become one of the key analytical techniques used to detect hidden explosives; a number of commercial systems is now available. These systems can be made rugged, hand portable and the analytical information can be presented in an easily understood format. A review of the application of IMS to the field of explosives detection is presented. This covers the detection of explosive in the vapour phase by using IMS "sniffers", and the detection of the low vapour pressure plastic explosives by sampling sub-nanogram traces of solid explosive and analysis in an IMS particulates detector. The methods by which IMS explosives screening will integrate into the security package of the future and some of the anticipated developments of IMS technology are discussed.

2. Background to Ion Mobility Spectroscopy (IMS).

Ion Mobility Spectrometry (IMS), once referred to as Plasma Chromatography, is an analytical technique which began its development in the late 1960s.[1,2] In any IMS system ionised gas molecules are separated according to their mobility as they drift through gas at atmospheric pressure whilst under the influence of an electric field. The principles of operation of IMS systems have been described thoroughly elsewhere,[1,3,6] therefore only a brief description of the construction and principles of operation will be presented here.

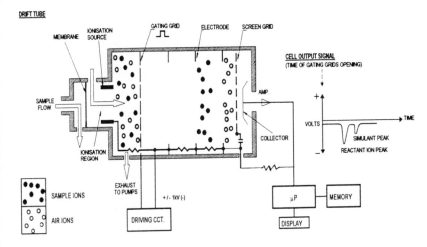

Figure 1. Schematic representation of IMS detector cell.

An IMS system is shown schematically in Figure 1.   The first requirement of an IMS system is a suitable ion source. The source is typically a beta emitter (most commonly Ni^{63}), this is a source of high energy electrons from which thermal electrons are created via frequent inelastic interactions with local gas molecules. During this process ions are created.

In a "dry" air system the ion species generated are pre-dominantly N_2^+ or O_2^+ positive ions and O_2^- negative ions. These ions will to varying degrees, cluster with water molecules, creating more complex cluster ions, depending on temperature and humidity levels. These ions are referred to as reactant ions. When molecules of the sample species are introduced into the ion source region then these molecules will interact with the reactant ions, and assuming they have a greater affinity for the associated charge, then charge transfer will take place creating ions of the sample species. Other mechanisms are possible to create characteristic product ions, such as the formation of adduct ions.

Other ion or electron sources have been used in conjunction with IMS systems and some will be described briefly later.

These ion species will then move toward the ion gate under the influence of an extracting electric field.   The polarity of the latter controls whether positive or negative ions are detected. The electronic gate, as first developed by Bradbury and Neilsen [4], consists of a two closely positioned sets of fine mesh wire. When the gate is in its closed state then a bias voltage is applied between the two sets of wires and a transverse electric field is created between alternate wires in the structure such that ions in the region of the gate will be attracted to the grid wires where they loose their charge and are prevented from entering the drift region.   When this bias voltage is pulsed off then a pulse of ions are allowed into the drift region. The duration of the pulse is typically a few 100μs; drift times are around 5 - 20ms.

In the drift region the sample ions move under the influence of the applied electric field. Due to the collisions between the sample ions and the drift gas molecules, separation takes place depending on the mobility of the sample ions. Ions with a higher mobility traverse the length of the drift region in a shorter time than ions with a lower mobility. The mobility of the ions will depend on their size, mass and shape. Hence, for each opening of the electronic gate, a pulse of electronic charge representing the different ion species in the sample will arrive at ion collector separated in time. This spectrum of ion current is sometimes referred to as a plasmagram. These current pulses can be easily measured, and fed into a microprocessor where signal processing takes place. For repeat gate cycles the signals can be averaged to improve the signal to noise ratio and analytical measurements of the output made.

The detailed theory of mobility has been treated elsewhere [1,5,6] and will not be presented here. It should be noted here, that the mobility of an ion species is defined by:

$$v_d = KE$$

where,

v_d = drift velocity,

E = electric field gradient,

and

K = the mobility of the species.

The mobility of an ion will be, to a first order approximation, a linear function of temperature and pressure. To generalise the mobility as a physical characteristic of the ion species, across a range of pressures and temperatures, the reduced mobility, K_o is defined, the first order contributions of pressure and temperature, being normalised out,

$$K_o = K(273/T)(P/760)$$

where,

T = absolute temperature ,

and

P = pressure in torr.

For known temperature and pressure characteristics in the IMS system sample ions can be identified through measurement of their drift time in a drift tube of known length and electric field.

3. The advantages of using IMS for explosives detection.

From the early days of the development of IMS its potential for the detection of explosive materials has been explored.[7]

The chemical characteristics of a large number of the compounds classified as explosives, their strong electron affinities, means that the efficiency with which they create negative ions through ion molecule reactions is high. Therefore the potential sensitivity of an IMS system to these compounds is high. Since the dimensions of a typical drift cell may be 30 to 50mm and that of an entire IMS system may be 200 to 300mm, it can be seen that IMS lends itself well to construction in a hand-held package.

The typical drift times of molecular species are tens of milliseconds, and therefore even with repeat spectral acquisitions and data averaging the output of analytical information can be on timescales under a second, giving near real time analysis. This defines the the main advantages of IMS; sensitivity, portability, autonomy and near real time analysis.

The restrictions on the applications of IMS technology are generally associated with the details of the ionisation processes and the limited resolution of an IMS detector. In single component studies then the ionisation processes are fairly well understood; however where field applications are concerned then multicomponent systems have to be considered, where other compounds may interfere with the ionisation of the target compound. This may lead, in the extreme case, to the detection of the target compounds being masked entirely by other compounds in the input "matrix", or more commonly the degree of quantification available from the output data being limited. The limited degree of resolution available can also mean that where a complex mix of species is introduced into the ion source then incorrect identification of background materials as a target compound may result.

Control over the design parameters of an IMS system which make it specific for the detection of explosive materials can to a great degree overcome these restrictions. This linked to the high sensitivity, and compatibility of systems to field deployment has meant that despite the qualitative nature of the output data there has been an extensive spread of IMS technology in the detection of hidden explosives, where often only a YES/NO response is required.

4. The important parameters in explosives detection by IMS.

Before some of the details of IMS based explosive detectors can be explained then it needs to be realised that the defining parameter in the design of a detector for hidden explosives is the vapour pressure of the target explosive. Where the target compounds are the nitrate based explosives NG, DNT, EGDN, where the typical vapour pressure at ambient temperatures is above 10^{-6} torr, then direct sampling of the target vapours can be used (i.e. the vapour levels are above 0.1 p.p.b. levels).

When such direct vapour sampling can be used then the advantages of using a non-porous dimethylsilicone membrane interface between the input vapour and the ion source have been shown [8]. This membrane will discriminate between target materials and interferents entering the source and thus reduce the interference in the ion source and therefore simplify the mobility spectrum.

The specificity for the detection of explosives by IMS systems has been improved by the use of specially selected reagent gases. The advantages of adding halide gases, typically chloride or bromide to the IMS system flow for the detection of explosives has been shown [9, 10 ,11]. The advantages can be seen by considering the ionisation processes. The dominant processes in the creation of negative ions are charge transfer and electron attachment through interaction with thermal electrons. The addition of reactant ions to the source region that have an electron affinity just below that of the target compound, reduce the number of interfering reactions as only species with a higher electron affinity than the reactant ions can be ionised readily and therefore detected, thus improving the selectivity of the system. Also reactant ions which can transfer charge to or cluster with the target compound can improve the efficiency of the creation of target ions due to the larger cross-section for interaction with an ion than with an electron, thus improving sensitivity. For the ultratrace detection of EGDN the preferential creation of the EGDN+Cl$^-$ cluster in a chlorine doped system over the EGDN+NO$_3^-$ in a nitrogen based system has been shown.[10]

The application of IMS systems to deployment in security screening for the detection of nitrate based explosives has been driven in the main by the following system characteristics:

1. The high sensitivity of IMS to the nitrate explosives EGDN, NG and DNT, providing minimum levels of detection in the sub p.p.b. levels. [12,13]
2. The near real-time response characteristics of the IMS giving output data on a second timescale.
3. Hand-portability of the systems coming in a suitcase size package or smaller.
4. Low running cost with very few system consumables.
5. Ease of operation, suitable for unskilled operators, the systems providing simple YES/NO type indicators.

This has meant that IMS systems will now be found in many of the world's airports providing security screening.

The success of such devices, often called "sniffers", has lead the international security communities to propose a program to tag the low vapour pressure plastic explosives with a higher vapour pressure component during manufacture. EGDN is one of the compounds proposed as a taggant.

5. Applications of IMS to the detection of plastic explosives.

With the security threat changing to plastic explosives in the late 1980s, so the developments of the technology had to respond.

Where the classification of target compounds are the low vapour pressure plastic explosives TNT, RDX, PETN, HMX etc. then the sampling mode must collect solid particulates of the target material before the vapour sample is thermally desorbed from the collected materials. This also demands that the IMS system is heated so that the desorbed vapours do not re-condense which would cause severe problems in the recovery of the system from sample input. In systems designed to detect the low vapour pressure plastic explosives the sample is taken by wiping the sampled surface collecting the explosive particulates, and from this wipe the sampled materials are thermally desorbed into a sealed pneumatic IMS system. The desorbed vapours will be entrained in a sample flow and transported to the ion source region. The description temperatures are typically 200°C with the detector kept at a some what lower temperature, typically 100°C, to prevent breakdown of some of the explosive. These elevated temperatures prevent the clustering which membrane sampling counteracts at low temperatures and this, together with the sealed pneumatics system, obviates the need for a membrane interface. At these elevated temperatures for some of the explosives (RDX, PETN, NG, for example) multi-component cluster ions can be formed [14]. The target ions combining with NO_3^- and Cl^- ions in the source are RDX+Cl^- and RDX+ NO_3^-; also larger cluster dimer ions can be formed (e.g. RDX_2Cl^-). The relative distributions of these ions are strongly concentration dependent. Figure 2 shows the IMS response typical of a few nanograms of RDX deposited on a sample wipe. From the low level spectrum, the RDX+Cl^- and RDX+ NO_3^- monomer responses can be clearly seen however the dimer . RDX_2Cl^- is only just detectable above the noise. As the input concentration increases the dimer peak would begin to dominate. These multi-component responses allow more specific detection algorithms to be utilised in the signal processing which search for the presence of at least 2 peaks.

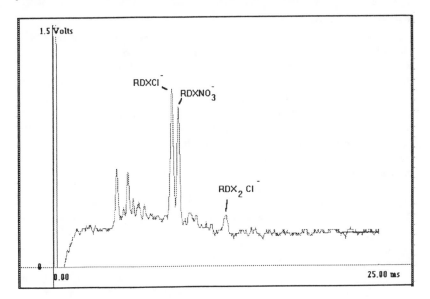

Figure 2. Typical IMS spectrum from nanogram levels of RDX.

Systems can again be made portable. However, the normal mode of operation would be such that they are bench mounted, since the sample filter would be taken to the surface to be sampled and the wipe taken, then brought back to the system where the analysis would be performed. A typical analysis cycle would take 10 to 15 s, a simple YES/NO screen again being provided. Several such systems are now commercially available. These give limits of detection in the low nanogram to sub-nanogram levels for RDX, PETN, TNT and NG, the materials most often associated with a terrorist threat. This level of detection will provide screening for hidden explosive devices from the traces of explosive residues deposited on the outside of the suspect device.

Probably as a consequence of the more complicated operational procedures and higher unit cost than for vapour detectors, the widespread usage of this technology is slow to grow. However, the continued threat from international terrorism, and the terrorist usage of plastic explosives is driving the increase in field deployment.

With its fast response times and simple data output, IMS lends itself well to automation. The development of sampling systems will lead to the interfacing of IMS detectors with the technologies already automated into security screening systems, X-ray systems and metal detectors. Such developments should lead to the fully automated security screening systems of the future[15].

6. Future developments in IMS technology, and their potential applications.

One of the major developments in IMS has been the development of a non-radioactive source. Developed primarily to combat the need for the regulatory controls when using radioactive source, it also has the promise of improved performance. The challenge to produce a highly reliable source with low power consumption, in a small package easily interfaced to an IMS system is considerable. These criteria have been achieved by using a dual point pulsed corona system. The corona pulse is synchronised to the opening of the ion gate, the pulsed nature of the corona extending the life of the corona point to in excess of 5000 hrs continuous operation. The source also generates a higher charge density than a Ni^{63} source giving the potential for higher sensitivities.

New IMS manufacturing methods have made further reductions in the size of IMS systems possible. The new "mini-IMS" system developments at Graseby make full IMS systems pocket book sized, (see figure 3) and at a unit system price that could make the more wide spread use of IMS detectors possible.

One of the major developments in IMS technology in recent years has been to interface an IMS system as the detector of another separation system. These have mainly focused on developments of IMS as a GC detector system; however some work as also investigated interfacing IMS and Liquid Chromatography[16]. Most developments to interface IMS systems to GC have been toward lab based systems, however hand-portable combined GC/IMS systems providing the two level separation in a fieldable package are available [17] (see figure 4) The additional degree of separation should allow applications where more complex samples cause problems to simple IMS systems. For example this may open the route to applications in contaminated site screening, reducing the need for expensive lab based analytical techniques.

Figure 3. A Graseby GI mini detector providing IMS detection in a pocket-book sized packages

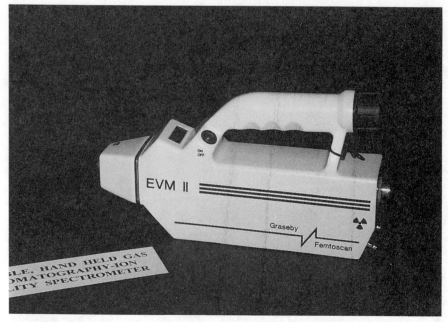

Figure 4   The EVMIII a hand-portable GC/IMS svstem,   designed for field screening applications.

7. References

1. Carr T. W. (Ed.) *"Plasma Chromatography"* Plenum Press: New York : *1984.*
2. Cohen M.J. Karasek F. W. J. *Chromatogr. Sci. 1970, 330.*
3. St. Louis R.H. Hill. H.H. *Critical Reviews in Anal.Chem.* 1990, 5,321.
4. Bradbury N. E. Nielson R.A. *Phys. Rev.* 1936, 49, 388.
5. Revercomb H.E. Mason E. A. *Anal. Chem.* 1975, 47, 970.
6. Eiceman G.A. Karpas Z. *"Ion Mobility Spectrometry"* CRC Press: 1994.
7. Cohen M. J. Wernlund R.F. Kindel R.C. *"Proceedings of the New Concepts Symposium and Workshop on Detection and Identification of Explosives"* 1978, 185.
8. Spangler G.E. Carrico J.P. *Int. J. Mass Spectrom. Ion Phys.* 1983, **52,** 267.
9. Spangler G.E. Carrico J.P. Kim S.H. *"Proceedings of the International Symposium on the Detection of Explosives"* 1983, 267.
10. Proctor C.J. Todd J.F.J. *Anal.Chem.*1984, **56,** 1794.
11. Lawrence A.H. Neudorfl P. *Anal.Chem.* 1988, **60,** 104.
12. Conrad F. *"Proceedings of the Third International Symposium on the Analysis and Detection of Explosives"* 1989, 1-1.

13. Danylewych-May L. Cumming C. *"Advances in the Detection and Analysis of Explosives."* Kluwer Academic Publ.: 1992, 385.

14. Fetterolf. D.D. *"Advances in the Detection and Analysis of Explosives."* Kluwer Academic Publ.: 1992, 117.

15. Hobbs J.R. Conde E.P. *"Advances in the Detection and Analysis of Explosives."* Kluwer Academic Publ.: 1992, 437.

16. Wittmer D. et al *Anal.Chem.*1994, **66**, 2348.

17. Snyder P.A. et al . *Anal.Chem.*1993, **65**, 299.

Security and Forensic Science as Applied to Modern Explosives

Christopher Ronay

PRESIDENT, INSTITUTE OF MAKERS OF EXPLOSIVES, WASHINGTON D.C., USA

When Alfred Nobel introduced his explosives to America in 1866, our lives were forever and for-the-better changed. With his inventions we were able to build more and better roads and waterways and accomplish massive national growth much faster than anyone could have imagined.

The explosives industry in America grew at a rapid and tumultuous pace through the end of the century until it became clear that the manufacturers needed an organization to set safety standards and assist the government in crafting the now burgeoning regulations governing the manufacture, transportation and storage of explosives.

The Insitute of Makers of Explosives was thus founded in 1913 to promote safety throughout the life of explosive products from manufacture to use. In the ensuing 83 years, our constant objective has been to provide technically accurate information to insure that industry safety standards are accurately and uniformly reflected in regulations.

Our member companies today produce more than 90% of the five billion pounds of commercial explosives consumed in the United States each year. And because no one knows more about explosives in America, we have been called upon repeatedly to assist law enforcement efforts in criminal investigations involving explosives.

No one has a greater interest than do we, in keeping explosives out of the hands of criminals. When the terrorist strikes with an explosive device, it hurts us all in many ways. It particularly impacts this industry because we are easily targeted by politicians looking for a quick fix in response to the hysteria that inevitably   follows a terrorist bombing.

The tragic and senseless criminal bombings of the World Trade Center in 1993 and the Oklahoma City Federal Building in 1995 certainly affected all of us. In each case, our politicians immediately sought to reassure their constituents that such events would never happen again. The Clinton administration and certain members of Congress reacted predictably by once again calling for an identification tagging program for commercial explosives to counter the terrorist bomber.

And, once again, they ignored established facts about bombing attacks. Terrorists do not often use commercial explosives. The most recent government statistics show that less than 1% of all criminal bombings, and none of the recent terrorist acts, employed commercial high explosive materials.

Nevertheless, our industry is again challenged with the politically satisfying tagging proposal and in response we have supported a comprehensive and objective study of the issues so that once and for all the questions about the efficacy of such a program can be answered publicly.

IME organized a coalition of industries representing explosives manufacturers, distributors, user industries, blasters and customer groups which voiced concerns about a hastily contrived explosives tagging program without a comprehensive study of all of the consequences.

The fact that looms largest, and bears repeating, is that terrorists do not use commercial explosives. It would seem prudent to first examine the merit of any identification tagging program in view of its limited value to investigators when such a small percentage of criminal bombings use commercial explosive charges.

Furthermore, placing taggants in all commercial explosives would confound crime scene investigations when in a short time taggants would be prevalent in our environment as so many products common in our daily lives are derived from commercial explosives.

The taggant which is historically considered for this purpose is the Microtaggant©, a tiny chip made of plastic and metal layers. These taggants are very expensive and would significantly raise the cost of mining operations and other commercial activities involving the necessary and legitimate use of explosives. Conservative estimates are in the area of 1.5 to 2 billion dollars annually.

Last April, Congress passed the *Antiterrorism Act of 1996.* This law calls for some significant enhancements for law enforcement agencies as they combat terrorism. It also ratified the ICAO Convention for the Marking of Plastic Explosives, which requires an aromatic detection agent in plastic and sheet explosive products to make them more "visible" to existing detection equipment.

Additionally, the Act requires a comprehensive study by the government, with independent oversight, of the following:

- ▸ The tagging of explosive materials for detection and identification;

- ▸ The feasibility of rendering common chemicals used to manufacture explosive materials, inert;

- ▸ The feasibility of imposing controls on certain precursor chemicals used to manufacture explosives; and

▸   Licensing requirements for the purchase and use of commercial high explosives.

Most important, the study must prove that taggants:

▸   Will not pose a risk to human life or safety;

▸   Will substantially assist law enforcement officers;

▸   Will not substantially impair the quality of the explosive materials ;

▸   Will not have a substantial adverse effect on the environment; and

▸   That the costs associated will not outweigh the benefits derived.

At the conclusion of the study, the U. S. Congress will further consider the advisability of a regulation on taggants. We consider the language of this legislation to be a victory for rationalism and a measured, practical response to a very charged emotional issue.

It is important in this effort to make it clear that an identification tagging program will not prevent bombings. I have been involved in explosives tagging research and study for 20 years, and based on my extensive experience as a bombing investigator, an identification tagging program is of very limited value to law enforcement. Criminals can readily circumvent a tagging program by stealing explosives, fraudulently purchasing explosives, making their own explosives, or using military explosives which would not be tagged.

I believe that it is irresponsible to squander valuable resources on something that is of limited value to investigators when we could put those resources to better use trying to prevent bombings through proactive investigative techniques and early detection. We should strive to detect the bombs before they can kill and maim.

Where the investigator is concerned, forensic science techniques and equipment have advanced over the years to where today explosives can be identified by their inherent constituents - their unique chemical fingerprints. Such technology can be useful when examining any explosive, homemade as well as commercially produced.

The instrumentation used in the analysis of bombing evidence includes the IMS, or Ion Mobility Spectrometer, which is used for screening specimens at the scene. Various types of mass spectrometers are used at the scene and in the laboratory to identify specific explosives.

One of the most useful pieces of equipment in the forensic laboratory is the Triple Stage Quadrapole Mass Spectrometer or TSQ. It is used to separate various ions in the specimen and identify the explosive by molecular mass. The most sophisticated instruments combine the technologies of gas chromatography and mass spectrometry. Called the GCMS, this equipment allows the forensic investigator to fingerprint the explosive in many cases, even when the explosive is homemade.

The other device components required to detonate the bomb are often of even greater forensic value. Electronic components and associated materials like containers and vehicles, in the case of vehicular bombs, can be identified and traced to the criminal. You will recall the investigation of the bombing of Pan Am Flight 103 and the forensic work that traced the tiny fragment of the electronic timer to the terrorists.

This work was done in forensic laboratories relying on the use of reference collections, product databases and intelligence programs. Taggants would have been of no use here, nor in any of the other high profile bombings we have experienced. As you know, the bombings of the World Trade Center and the Oklahoma City Federal Building involved homemade explosives which would not contain any taggants. So it is ironic that the political rhetoric calling for taggants seems to be motivated by crimes that it does not propose to address.

Often, I hear the argument that the explosive tagging program in Switzerland is an example of the successful use of taggants in explosives. Those who offer this example are not familiar with that program and its lack of specificity. I wonder if the Swiss explosive manufacturers would be as tolerant of a program that required three daily taggant changes allowing no contamination between batches, and thus, no reworked product. And yet, we must contend with this program which is held out as an example of what can be done to stem terrorism.

The April anti-terrorism legislation excluded any study of tagging black and smokeless powders, even though these valuable and useful products are used in a much greater percentage of bombs than commercial explosives. However, following the bombing at the Atlanta Olympic Games this summer (pipe bombs filled with smokeless powder) additional legislation was passed which does mandate a study of tagging black and smokeless powder. More important, $18 million was appropriated to conduct these studies and a requirement was added to examine methods to enhance prevention and detection technologies available.

The Department of Treasury, responsible for these studies, must issue a report in April 1997. It will likely state that further research needs to be done before any recommendations can be made.

In the mean time, IME has recommended that we tighten up the security loopholes in our explosives laws. We have long recommended a national licensing program to require a federal permit to purchase and possess explosives in the U.S. Remarkably, only one half of our states have such a requirement at present. And, we continue to promote the development of detection and prevention technologies to keep the bomb from going off in the first place.

In point of fact, industry trends toward bulk delivery, nonelectric initiation systems and products which are less useful to criminals have resulted in an annual reduction of commercial high explosives involved in criminal cases. Alfred Nobel would be pleased with these developments, I'm sure.

In conclusion, IME believes that it is only through enhanced security measures, innovations in explosives detection, and the further development of forensic science technologies, that we can hope to limit the atrocities of future terrorist bombers. And in doing so we may usher into the Twenty-first Century, the inventions of Alfred Nobel with the accolades and appreciation they deserve

Dag Hammerarskjöld once wrote, "It was Alfred Nobel's dream that through the advancement of science, conditions will be created for a better life and for peace among all people."

So, let us praise the advancing science of explosives in creating a better life for all. And let us abhor the destruction of the terrorist as a threat to peace among all people.

Session 4. Explosives Design and Bulk Explosives

Descriptions of Condensed Phase Detonations Based on Chemical Thermodynamics

M. Braithwaite[1] and W. Byers Brown[2]

[1] ICI RESEARCH AND TECHNICAL CENTRE, WILTON, CLEVELAND TS90 8JE, UK
[2] UNIVERSITY OF MANCHESTER, MANCHESTER M13 9PL, UK

1 INTRODUCTION

Theoretical studies of the detonation process began in the 19th century about twenty years after the development of dynamite. Much of the understanding of detonation phenomena has been based on research in gas explosions, beginning with the early work of Berthelot[1] (1882) and Dixon[2] (1893) who attributed the high detonation wave velocities to properties of the explosion product molecules. Schuster[3] (1893) was the first to point out similarities between detonations and unreactive shock waves and to suggest the now accepted view that detonation is a shock wave supported by chemical reaction.

The first attempts at a thermodynamic analysis were independently reported by Chapman[4] (1899), Jouguet[5] (1903) and Michelson[6] (1891). In CJ (Chapman-Jouguet) theory, the detonation process is treated as a 1-D wave of infinitesimal thickness and without mass, momentum or energy loss in an inviscid fluid (Figure 1). Detonation velocities could be calculated as the minimum subject to these constraints and were largely governed by the thermodynamics of the product molecules.

This simple CJ model has achieved remarkable success over the past hundred years in predicting detonation velocities where lateral losses have been small and chemical/diffusion rates fast. Refinements have come in improved descriptions of the thermodynamic properties of the multi–component multi–phase mixtures that comprise the detonation products from condensed phase explosions.

Zeldovich[7] (1940), von Neumann[8] (1942) and Doring[9] (1943) extended the CJ model allowing irreversible reaction in a finite reaction zone behind a shock jump discontinuity but still subject to the CJ model constraints. At the same time Devonshire[10] (1943) found the more general criterion not restricted to 1–D detonations pointing out that the energy release by chemical reaction has to be balanced by the energy loss by radial expansion at the sonic point.

Real detonations are known to have the following properties: (1) curved shock front (2) lateral losses (3) cell structure (4) complex finite rate and diffusion processes (5) ability to spin (6) transport effects including turbulence. Modern hydrodynamic theories have taken account of some of these phenomena,(1),(2), (6), to account for the effect of charge diameter and confinement and consequent flow divergence effects (eg

Wood and Kirkwood[11] (1954), Bdzil[12] (1981), Kirby and Leiper[13] (1985)). The advent of modern powerful computers has enabled the use of finite element calculations based on individual phenomenological models to look at detonations in different geometries (eg Tarver[14] (1981)). With the recent work on exact solutions for cylindrical steady state detonations (Cowperthwaite[15]) there are now opportunities to further validate the more general numerical computer models.

Other theoretical models involving chemical reaction breakdown and shock discontinuity zones have been introduced (Dremin[16] (1994)) controlling detonation stability and failure, and the onset of the detonation process respectively.

No one approach has been capable of capturing the best available thermodynamics, hydrodynamics and chemical kinetic/ diffusion model. The work described in this paper is concerned with a description of the detonation process making the maximum use of information derived from the chemical thermodynamics of the products and unreacted explosive and developing a framework for including reaction and dissipative processes in terms of a global maximum entropy criterion.

1.1 Prediction of the Performance of Commercial Explosives

The ideal detonation characteristics of an explosive provide an estimate of the maximum performance that can be achieved based on a particular explosive formulation and initial density. This predicted performance is largely governed by the equations of state (EoS) used to describe the different products and phases produced.

Estimates of performance, including velocity of detonation (VoD), energy release, detonation pressures and particle velocities will be at variance with most experimental data for commercial explosives due to lateral losses and finite rate chemistry in these systems. These calculational procedures can, however, be validated against military (HE) explosives where these losses are small: in many circumstances the EoS of individual product species can also be compared with experimental compression data or for small systems with ab initio calculations.

Results from the predictions of an ideal detonation code can provide the following information for use in other non-ideal detonation or hydrodynamic codes:

- an estimate of the maximum obtainable velocity of detonation,

- a first estimate of the energy split and the total energy release,

- a database for simpler EoS parameterisation.

There are a large number of different ideal detonation codes in existence. The differences in the predictions of these codes are almost entirely due to the different fluid and solid product EoS used.

2 IDEAL DETONATION

An ideal detonation is illustrated in Figure 1. It consists of a one–dimensional process where mass, momentum and energy are conserved whilst thermal, mechanical and chemical equilibrium are attained in the end products. The ideal (CJ or Chapman–

Jouguet) detonation criterion further requires that the VoD is a minimum subject to the above constraints.

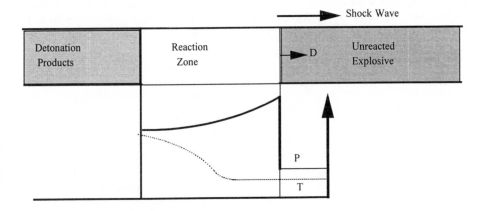

Figure 1 *Ideal (ZND) Detonation*

In addition to the product EoS, an ideal detonation calculation requires only a knowledge of the explosive composition (in terms of elemental makeup), its initial density and heat of formation.

An ideal detonation prediction therefore ignores the following known physics:

- CJ assumption: most observed detonations probably lie on the *weak* branch,

- 1–D planar structure: cell structure, flow divergence and shock front curvature are known to occur, albeit to only a limited extent in HE systems,

- attainment of chemical equilibrium: it is likely that there are both diffusional and chemical kinetic constraints to achieving full conversion to chemical equilibrium.

The above assumptions limit ideal detonation codes in their application in commercial explosives to:

1. ideal (infinite diameter) VoDs: CJ state conditions,

2. equilibrium expansions (isentropes, isotherms, isobars) from a starting state at equilibrium and energy releases during expansions.

This excludes calculations where there is a chemical kinetic component or loss term (VoD vs diameter effects, explosion fume, shock and heave energy partitioning): where these calculations have been attempted with ideal detonation codes it has only been accomplished by resort to empirical fixes which cannot be generally extrapolated.

3   EQUATIONS OF STATE

The thermodynamics of all the product species are defined in terms of their EoS for each phase in which they may be present. For the conditions that prevail in a condensed phase detonation it is normally sufficient to calculate the intra–molecular properties of a single molecule and inter–molecular forces between molecules separately, describing the former as an ideal (low pressure) system. It is therefore assumed that there is no change in molecular vibration or rotation due to high pressure.

The thermodynamic parameters such as energies, entropies and pressure can be determined from a sum of the ideal and non–ideal fluid contributions. The former quantities can be calculated using fitted representations of standard ideal gas tables for the temperature range of interest.

The non-ideality of the products (due to intermolecular forces) is described by various EoS, one for each phase present. These EoS have been chosen on the basis of their ability to represent the repulsive intermolecular forces which will dominate at high pressures. Over the past 50 years a wide range of fluid and solid EoS have been used for detonation products. Explosives performance in commercial products is largely governed by the fluid phase EoS: those in common use include Empirical EoS: BKW (Becker-Kistiakowsky-Wilson) (Fickett and Davis[17] 1979), Semi-Empirical EoS: JCZ3 (Jacobs-Cowperthwaite-Zwisler) (Cowperthwaite and Zwisler[18] 1976) and Fundamental EoS: Intermolecular Potential Based:WCA (Weeks-Chandler-Anderson) (Chirat and Pitton–Rossillon[19] 1981).The preferred EoS for any detonation calculation must be that which provides the most realistic representation of the thermodynamics of the product species. This can be established based on the following:

- agreement with shock Hugoniot data for pure unreactive species – pressures (and temperatures, where available),

- self consistency in EoS parameters used – in accord with expectations based on critical properties,

- incorporation of appropriate chemistry – polar terms for polar molecules etc.,

- well behaved EoS and its derivatives eg. adiabatic, Gruneisen gammas etc.,

- correct high and low pressure asymptotic behaviour,

- reasonable agreement between measured and predicted VoD's for HE's.

Advances in theoretical EoS for fluids based on both perturbational and variational approaches have resulted in a number of EoS available for high pressure fluids based on the fundamental equations of statistical mechanics (Kang[20] et al 1985). The predictions of these different fundamental EoS are similar since the equations themselves can be validated against Monte–Carlo simulations with the intermolecular potential of interest.

The major drawback of ideal detonation calculations using theoretical EoS has been the large amount of computer time required to calculate the excess thermodynamic contributions. In general terms the Helmholtz Free Energy, A, for a perturbational approach, is expressed by,

$$(3.1)$$

$$A = A_i + A_{excess}$$

where A_{excess}, the intermolecular contribution to the energy is given by,

$$A_{excess} = A_{HS} + A_1 + higher\ order\ terms\ldots \qquad (3.2)$$

where the hard sphere (HS) contribution is determined by judicious choice of hard sphere diameter minimising the higher order term contributions.

For a mixture of detonation products, standard mixture and combination rules are applied. Corrections to the normally repulsive forces can be readily made to allow for the attractive force contribution from polar molecules such as H_2O, NH_3 etc. Intermolecular parameters used can be obtained from molecular beam scattering studies, ab initio calculation, by fitting to high pressure shock Hugoniot or static data or even by resort to Corresponding States criteria based on known critical properties.

Calculation of ideal detonation characteristics, formally solving all the statistical mechanical equations, requires substantial computational resources and is beyond the means of small desktop and portable computers. This difficulty can be circumvented by using an analytic representation of the fluid EoS (Byers Brown[21] 1987). This is achieved in two stages:

1. choice of *canonical* variables
   - the adoption of variables such that the thermodynamic excess properties vary more smoothly in the P, V, T region of interest,

2. appropriate polynomial fit
   - provision of optimum interpolation procedure.

It is now possible to perform complete ideal detonation calculations in a matter of a minute on a 486 series PC or equivalent.

3.1 Comparison of equation of state predictions and Hugoniot data

Shock Hugoniot data are available for a number of pure, non-reactive materials. For a high pressure EoS to have any credence there must be a reasonable correspondence between predicted and experimental shock Hugoniot data. In some instances, shock temperature information is available for some species.

For the case of the major common explosive detonation products, these data (both velocities and temperatures) have been published for both H_2O and N_2 (Byers Brown and Braithwaite[22] 1990). These data are compared with Hugoniot predictions for the three EoS described here (Figures 2 to 5).

For the case of the Nitrogen Hugoniot, the WCA EoS clearly fits the experimental shock Hugoniot well: the other EoS lie on the periphery of the P,V data. Comparison with measured shock temperatures indicate both the WCA and JCZ3 are in reasonable agreement over a wide range of conditions. For the case of the polar water system it is clear that only the WCA EoS both adequately fits the water Hugoniot and is in reasonable agreement with the published shock temperature data: as the WCA EoS in IDeX contains a polar term this was only to be expected.

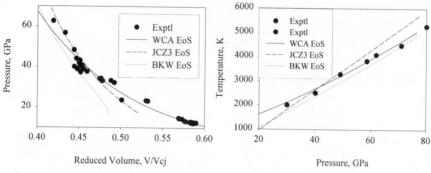

Figure 2 *Water Hugoniot: shock pressure*   Figure 3 *Water Hugoniot: shock temperature*

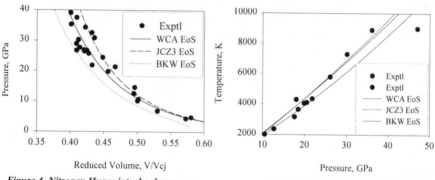

Figure 4 *Nitrogen Hugoniot: shock pressure*

Figure 5 *Nitrogen Hugoniot: shock temperature*

4   IDEAL DETONATION CODES

All ideal detonation computer codes essentially comprise   a means of calculating chemical equilibrium in a multi–phase multi–component non–ideal mixture subject to a variety of constraints or conditions.

The code consists of:

- user interface (via Windows or similar),

- databases - explosives ingredients, product species thermodynamic parameters, results,

- chemical equilibrium solver: constrained minimisation of Helmholtz Free energy,

- Equation of State subroutines (each phase),

- output files and graphics interface,

- Help facility.

Other codes have similar facilities but, in addition to EoS used and product species and phases included, will differ in the method of calculating chemical equilibrium: this can have implications for code robustness. With the advent of faster small computers and more efficient and flexible compilers the majority of ideal detonation codes can reside on a portable machine.

4.1 Comparison of code predictions

In addition to the validation of the EoS discussed earlier there are two further criteria for discriminating between the predictions of different ideal detonation codes:

1. Comparison of predicted detonation velocities (monomolecular explosives - experimental diameters much larger than critical) with those measured experimentally. Detonation velocities can be measured to accuracies less than 1% and for the case of military explosives, it is generally accepted that measured velocities should correspond to Chapman-Jouguet estimates (Mader[23] 1979). A comparison of predictions for a range of these explosives with the three fluid equations of state is given in Table 1 using standard BKW and JCZ3 parameter sets (Cowperthwaite and Zwisler[18] 1976: Byers Brown and Amaee[24] 1991) and ICI's IDeX (Freeman[25] et al 1990).

Table 1 *Comparison of Predicted and Experimental Velocities of Detonation*

| Explosive | Initial Density g/cc | VoD, m/sec | | |
		Expt.	WCA	JCZ3	BKW
HMX	1.189	6713	6741	6651	6381
NG	1.60	7700	7650	7531	7712
NM	1.128	6290	6283	6110	6351
PETN	1.770	8295	8247	8213	8446
PETN	0.95	5330	5360	5531	5701
RDX	1.00	6100	5981	6094	6143
TATB	1.895	7860	8133	8071	8118
TETRYL	1.710	7850	7751	7602	7651
TNT(s)	1.64	6950	6896	6279	7137
TNT(l)	1.45	6590	6489	6319	6570

2. Behaviour of isentrope slopes - adiabatic gamma.
   For detonation product media at modest densities and in the absence of phase transitions, the slope of the isentrope, γ where,

$$\gamma = -\frac{V}{P}\left(\frac{\partial P}{\partial V}\right)_S \qquad (4.1)$$

would be expected to be monotonic i.e. with no turning points or inflections.

A comparison of the three EoS predictions for PETN, detonating from its maximum crystal density of 1.77 g/cc, is given in Figure 6 alongside a JWL EoS (see later) fit. It is clear that both the JWL and JCZ3 EoS give characteristic turning points, well below CJ densities. The WCA EoS at all but the highest density is monotonic: the decrease in γ at the high density can be attributed to the formation of solid Carbon.

5 SIMPLIFIED REPRESENTATIONS

Ideal detonation estimates of the detonation velocities of commercial explosives can only predict a maximum attainable velocity: the predictions take no account of finite chemical rates or energy dissipation terms. Where a more complete study of the fluid dynamic characteristics of a non-ideal detonation process is undertaken it is not practical to incorporate a full EoS and simplified representations, chemistry implicit, such as the JWL and Williamsburg EoS are used. In the selection of a simplified EoS it is desirable to include properties such as:

1. simple analytic form, readily fitted to principal isentrope,

2. statistical mechanical basis,

3. correct asymptotic behaviour,

4. complete EoS (include Pressure, Volume, Temperature, Energy).

5.1 JWL EoS

The JWL EoS(Fickett and Davis[17] 1979) has been extensively used by explosives engineers to describe isentropic expansion of detonation products. It is an entirely empirical (and incomplete) EoS which, for an isentrope, has the form:

$$P = a_1 e^{-R_1 V} + a_2 e^{-R_2 V} + \frac{a_3}{V^{1+\omega}} \tag{5.1}$$

where a_1, a_2, a_3, R_1, R_2, and ω are constants and $V = v/v_0$: P, v denote pressure and volume. An expression for energy is readily obtained by integrating this expression. The parameter ω can be shown to be equal to the Gruneisen gamma coefficient, $1/V(\partial E/\partial P)_V$: the equation is therefore of a constant Gruneisen Gamma form and this is a poor approximation for a quantity that can vary by more than a factor of two. The arbitrary form of this equation also gives anomalies in the adiabatic gamma plot. It therefore satisfies only (1, 3 and 4) of the above criteria.

5.2 Williamsburg EoS

The Williamsburg EoS (Byers Brown and Braithwaite[27] 1991) for an isentrope is given by,

$$E = \frac{PV}{g - 1} \tag{5.2}$$

where g is given by,

$$g = \gamma_0 + \sum_{k=1}^{N} \frac{\gamma_k}{1 + \beta_k v} \tag{5.3}$$

where γ_k and β_k are constants: the order of approximation, N, is less than or equal to 4. Were g a constant this would reduce to a polytropic form. This equation can be shown to satisfy all the prerequisites listed earlier with little penalty in terms of complexity.

5.3 Application to an IDeX Isentrope

The calculated isentrope for a typical commercial explosive[26](Figure 7) has been fitted by both the Williamsburg and JWL EoS. The former EoS, whilst constrained to match CJ conditions and isentrope gradient has much smaller errors than JCZ3.

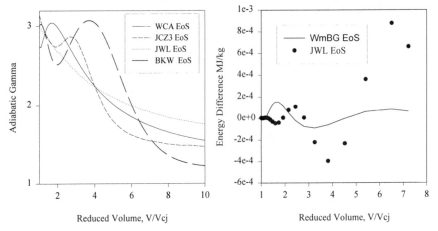

Figure 6 *Adiabatic gamma plots*

Figure 7 *Errors in energy prediction along isentrope*

6 MAXIMUM ENTROPY OF REACTION CRITERION FOR STEADY DETONATION

All the foregoing sections have dealt with ideal detonation. We shall outine a new approach to the understanding and calculation of some of the non-ideal effects mentioned in the introduction.

In the steady detonation process the total entropy of a given mass of explosive increases enormously, as one would expect. However, the fact that it is actually a minimum at the CJ state with respect to states with greater detonation velocity has always seemed puzzling and contrary to intuition. Chapman was happy in the erroneous belief that it was indeed a maximum, thus strongly supporting his choice of the minimum detonation velocity: a by-product of Jouguet's independent discovery of the CJ condition was his proof that, on the contrary, the entropy is a minimum.

Although detonation is a phenomenon which takes place in isolation from its surroundings, since it is not in thermodynamic equilibrium, there is no need for the entropy to be a maximum. The seemingly odd behaviour of the entropy can be made more understandable by adopting the point of view of ZND theory mentioned earlier. According to this theory the process can be divided into two distinct phases: first, a shock wave compresses the explosive in a very thin layer to a high pressure extremely quickly before any decomposition occurs: second, the decomposition reaction begins as a result of the high shock temperature and pressure and continues to completion in the reaction zone which drives the shock wave (detonation driving zone). The entropy change, ΔS can therefore be written as the sum of two terms: that due to the shock wave (W) and that due to the subsequent reaction (R),

$$\Delta S = \Delta_W S + \Delta_R S.$$

Even though irreversible chemical reactions are taking place, it would be very odd if the second term, the entropy of reaction $\Delta_R S$, were not a maximum at the end of the detonation process. This is because ever since the days of J W Gibbs even irreversible reactions have been treated by equilibrium chemical thermodynamics. This is achieved by the simple device of breaking the process down into infinitesimal steps between which the system is in thermodynamic equilibrium except for the reaction coordinate. Each step can then be considered to be an infinitesimal change from one equilibrium state to another.

It is not difficult to see that $\Delta_R S$ is indeed a maximum for one-dimensional detonation. Figure 8 is the familiar P–V diagram showing the Rankine-Hugoniot (RH) curves for the unreacted explosive and the fully reacted products. $\Delta_W S$ is the entropy change on going from the initial state to a state F at the back of the shock wave, and $\Delta_R S$ is the change on going down the Rayleigh line from F to a possible final state B somewhere on the fully reacted RH. The behaviour of the total entropy is shown in Figure 9. The entropy of reaction is the difference between the shallow product curve well and the deeper explosive curve well. This difference is sketched in Figure 10, and it is easy to see why it is a maximum at the CJ state.

It is natural to conjecture that $\Delta_R S$ will also be a maximum at the end of the sub-sonic detonation driving zone in more than one dimension. The situation for the two-dimensional (2D) case of a cylindrical charge is sketched in Figure 11. Note the curvature of the shock front, the divergence of the flow indicated by curved streamlines, and the so-called sonic surface which marks the end of the region which drives the detonation and which always occurs just before reaction is complete. The partial differentiation equations describing steady 2D detonation based on the conservation of mass, momentum and energy, and a chemical rate law, have been studied for 50 years, but their solution has been extremely difficult. This is mainly because the steady VoD and the shape of the shock front and the rest of the detonation driving zone are among the quantities to be determined by the solution of the reactive flow equations. Although a principle of maximum entropy of reaction does not contribute much that is new in the 1D case, since it is in essential agreement with C–J and ZND theories, it could play a useful role in the multi-dimensional cases. This is because it allows a variational approach of successive approximations to the calculation of the steady VoD and shape of the detonation driving zone. Trial forms of the flow and shock-front shape involving variational parameters can be introduced, and their values found by maximizing the entropy of reaction subject to the conservation laws.

A simple treatment of this kind has been carried out and gave very promising results (Byers Brown[28] 1996). Among the appealing features of a robust variational treatment of this kind are the possibility of using realistic EoS's such as the Williamsburg EoS discussed earlier, and realistic chemical kinetic laws, and extensions to calculate the effects of confinement, which are important in actual applications.

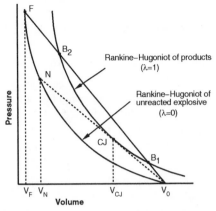

Figure 8 *P–V Diagram ZND Detonation*

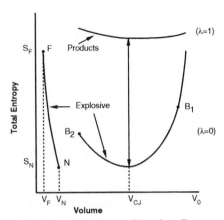

Figure 9 *Explosive and Product Entropy*

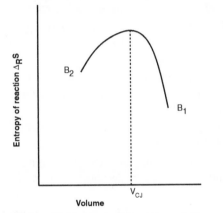

Figure 10 *Entropy of Reaction – 1 D case*

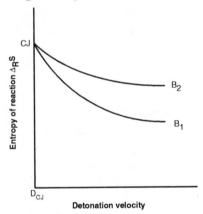

Figure 11 *Entropy of Reaction – 1 D case*

REFERENCES

1. M. Berthelot, *Compt. rend.* 1882 **94**.

2. H. B.Dixon, *Phil. Trans.* 1893 **184**.

3. A. Schuster, *Phil. Trans.* 1893 **184**.

4. D. L. Chapman, *Philos. Mag.* 1899 **47**.

5. E. Jouguet, *Mathem. J.* 1905.

6. V. G.Michelson, *Utshenie Zapiski Imperatorskogo Moskovskogo Universiteta, Moskva,* 1891 **10**.

7. Ya. B. Zeldovich, On the Theory of the Propagation of Detonation in gaseous systems, *Zh. Eksp. Teor. Fiz.* 1940 **10**.

8. J. von Neumann, Theory of Detonation Waves, *OD-02 NOSRD* 1942.

9. W. Doring, On Detonation Processes in Gases, *Ann. Phys.* 1943 **43**.

10. A. F. Devonshire, Theoretical Research Report, MOD 1943 **3**.

11. W. W. Wood & J. G. Kirkwood, Diameter Effect in Condensed Explosives *J. Chem. Phys.* 1954 **22**.

12. J. B. Bdzil, Steady State Two-Dimensional Detonation, *J. Fluid Mech.* 1981 **108**.

13. I. J. Kirby & G. A. Leiper, A Small Divergent Detonation Theory for Inter-molecular Explosives, *8th Detonation Symp. (Intnl)*,1986 NSWC MP 86-194.

14. C. M. Tarver, Modeling Two-Dimensional Shock Initiation and Detonation Wave Phenomena in PBX 9404 and LX-17, *7th Detonation Symp. (Intnl)*, 1981 NSWC MP 82-334.

15. M. Cowperthwaite, An Exact Solution for Axial Flow in Cylindrically Symmetric, Steady-State Detonations, *Phys. Fluids* 1994 **6**.

16. A. N. Dremin, Towards Detonation Theory *J Physique IV*1995 **C4:259**.

17. Fickett W. & W.C. Davis, *Detonation*, Univ. California Press 1979.

18. Cowperthwaite M. & W.H. Zwisler, The JCZ equations of state for detonation products and their incorporation into the TIGER code. *6th Detonation Symp. (Intnl.)*, 1981 162, ACR-221.

19. Chirat R. & G. Pittion–Rossillon, Detonation properties of condensed explosives calculated with an equation of state based on intermolecular potentials. *6th Detonation Symp. (Intnl.)*,1981 NSWC MP 82-334.

20. Kang H. S., Lee C.S, Ree T., Ree F. H., A perturbation theory of classical equilibrium fluids. *J. Chem. Phys.* 1985 **82(1)**.

21. W. Byers Brown, Analytical representation of the excess thermodynamic equation of state for classical fluid mixtures of molecules. *J. Chem. Phys.* 1987 **87**.

22. W. Byers Brown & M. Braithwaite, Sensitivities of adiabatic and gruneisen gammas to errors in molecular properties of detonation products *9th Detonation Symp. (Intnl.)*,OCNR 113291-7, 1990.

23. C L Mader, *Numerical Modeling of Detonations*, Univ. California Press 1979.

24. W. Byers Brown & B. Amaee, Review of equations of state of fluids valid to high density, *HSE Contract Research Report*, RPG2518 1991.

25. T. L. Freeman, I. Gladwell, M. Braithwaite, W. Byers Brown, P.M. Lynch & I.B. Parker, Modular software for modelling of detonation of explosives. *Math. Engng. Ind.*1990 **3**.

26. M. Braithwaite, W. Byers Brown & A. Minchinton, The use of ideal detonation codes in blast modelling, *Rock Fragmentation by Blasting*, 1996 Balkema.

27. W. Byers Brown & M. Braithwaite, Analytical representation of the adiabatic equation for detonation products based on statistical mechanics and intermolecular forces. *Shock Compression of Condensed Matter*, Elsevier, 1991.

28. W. Byers Brown, Maximum Entropy of Reaction Criteria for Steady Detonation, *International Conference of Shock Waves in Condensed Matter*, 1996 St Petersburg.

Bulk Explosives

W. Mather

ICI EXPLOSIVES EUROPE, ROBURITE CENTRE, SHEVINGTON, WIGAN, LANCASHIRE
WN6 8HT, UK

1. INTRODUCTION

The first bulk explosive was black powder but its use declined
following the introduction of nitro-glycerine based products in the 19th
century to such an extent that in extractive terms it finds only limited
application in slate blasting and monumental stone blasting, and even
here it is slowly being replaced by other techniques.

2. ANFO

ANFO was known about in the latter part of the 19th century but it came to
prominence as a potentially low-cost, powerful explosive following the
introduction of porous prills in the 1940's and the Texas City catastrophe in
which a large quantity of coated ammonium nitrate (AN) accidentally
detonated. It was in 1956 that the first oxygen balanced ANFO was
developed and field trialled by the Iron Ore company of Canada and
Canadian Industries Limited, now known as ICI Canada and part of the ICI
Explosives business. ANFO had a number of advantages over the packaged
explosives then available. They were:-

1. *Low cost*
2. *Ease of manufacture*
3. *In bulk form it fills the complete cross-section of shothole and provides 100% coupling with rock, an important performance benefit.*
4. *Easy to handle and load*

5.   *Problems of cartridge separation leading to partial misfires were avoided. This provided consistent performance and eliminated unfired cartridges in rockpiles, an important safety benefit.*

These advantages are still apparent today and they explain why ANFO remains so popular.

1960, or thereabouts, saw the first on-site automatic mixing and loading of ANFO using bulk trucks (figure 1). Their use enabled manpower to be cut and at the same time allowed larger blasts to be fired, thereby reducing the down-time wasted in clearing sites prior to blasting. The financial benefits are obvious.

The disadvantages of ANFO were significant, as detailed below, but not sufficient to outweigh the advantages. They were:-

1.   *Little or no water resistance*
2.   *Low density and thus relatively low bulk strength which limited its application as a base charge in quarrying, civil engineering applications, etc.*
3.   *Low VOD and thus detonation pressure which exaggerated the limitations described in 2. above*

The disadvantages or limitations of ANFO led directly to the development of slurries.

3. SLURRY

Slurry is a saturated aqueous oxidiser solution (largely AN) in which are dispersed excess solid oxidants (e.g. porous prills of AN) and a sensitising fuel. The fuel may be explosive (e.g. TNT) or non-explosive (e.g. aluminium), but it must ensure stable propagation of detonation when initiated by a suitable primer.

Slurries are made water-resistant by thickening and cross-linking the liquid phase to prevent water moving into (or out of) the product.

ANFO TRUCK

FIGURE 1

 **Explosives**

Slurries are almost completely insensitive to accidental initiation by shock, impact and friction and they overcome the main disadvantages of ANFO in that they have:-

1.   *Good water resistance*
2.   *Relatively high density, and thus high bulk strength when compared with ANFO and conventional packaged explosives*

They are also available in a range of strengths and different compositions are available to meet the requirements of high strength base charges and weaker column charges.

The high bulk strength of bulk slurries compared with bulk ANFO and packaged explosives enables increased burdens and spacings to be employed.

The first slurry truck, which blended and loaded slurry into shotholes, was introduced in the early 1960's. All the component semi-products were non-explosive and only when they were mixed and delivered into shotholes was an explosive formed. This process revolutionised the explosives industry in that it improved safety and made available low cost explosives that could be used in wet holes.

The improved safety occurred in both manufacture and use. It also offered potential for significantly reducing the quantity of explosive transported on public roads.

It is reported by Cook [1] that the availability of low cost explosive contributed to an expansion of the explosives industry in that it improved the economic viability of certain opencast mining operations to such an extent that it allowed them to proceed, whereas ordinarily they may not.

Slurry trucks (figure 2) are equipped to automatically mix and load into shotholes, using push button control, two or more strengths of slurry, either singly or in combination and without interruption, e.g. a strong base charge followed immediately by a weaker column charge. They can operate in a wide range of temperatures and load product into both wet and dry shotholes.

SLURRY TRUCK

FIGURE 2

 Explosives

Whilst the development of bulk slurry represented a real step forward in bulk explosive technology, the search for even lower cost products that were easier to manufacture and deliver into shotholes resulted in the development of emulsion explosives.

4. EMULSION

Emulsion explosives were first introduced into the marketplace in the late 1960's by Atlas Powder Company, now part of ICI Explosives.

Bulk emulsions require no chemical sensitisers and are essentially entrapped air within a water-in-oil emulsion of oxidiser solution (primarily AN) suspended as microscopically fine droplets surrounded by a continuous fuel phase (primarily oils and waxes). The emulsion is stabilised against liquid separation by the addition of an emulsifying agent. The entrapped air may be in the form of ultra-fine air bubbles or air entrapped in glass micro-balloons, etc.

The intimate mix of oxidiser and fuel resulting from particle sizes of typically 1 micron produces compositions with higher VOD's and detonation pressures than slurries. Since the energy partition is heavily biased towards shock energy at the expense of heave energy, the emulsion phase is almost always blended with AN when used in the surface extractive industries, the proportion of AN depending upon the energy partition required and whether or not the shotholes are wet. For example, in strong igneous formations where intense shock is required for fragmentation purposes blends with less AN may be used than in weaker limestone formations.

The type of loading-out equipment also affects the percentage AN in the blend, e.g. where a front-end loader is employed (which cannot dig very well) a looser rockpile is required than with a power shovel. Thus more AN would be used with a front-end loader to increase the heave energy.

In wet holes water resistant product is required, typically containing up to 40% AN, whereas in dry holes augured product typically containing 70% AN may be used.

The advantages of bulk emulsion compared with bulk slurry are broadly as follows:-

1. *Increased VOD and detonation pressure*
2. *Improved water resistance*
3. *Wide ranging flexibility in varying energy output to meet the demands of all blasting situations*
4. *Quicker, slicker, cleaner and greener operation*

Emulsion trucks are available to blend and deliver any composition between heavy ANFO (figure 3) which typically contains 30% emulsion phase and pumped products typically containing 70% emulsion phase. This flexibility allows the mining or blasting engineer to select the composition most suited to the needs of a particular blasting proposition. The actual composition chosen will depend upon a whole range of criteria, e.g:-

1. *Wet or dry holes*
2. *Sleep time in shotholes*
3. *Rock formation*
4. *Operation, e.g. quarrying or opencast coal mining*
5. *Loading out equipment*
6. *Burden and spacing*

Different blends can be used one after the other in different sections of the same hole and without interruption, to provide complete flexibility in blast design.

5. SURFACE APPLICATION OF BULK EXPLOSIVES

ANFO can be mixed and loaded into shotholes either by hand or automatically using bulk trucks which are almost exclusively site specific, neither process having changed much over the years.

Bulk slurries have largely been replaced by bulk emulsions and a number of different delivery units and chassis are available to suit different applications.

Delivery units are available in a variety of capacities. Bin capacities can also be varied to carry different proportions of AN and emulsion phase.

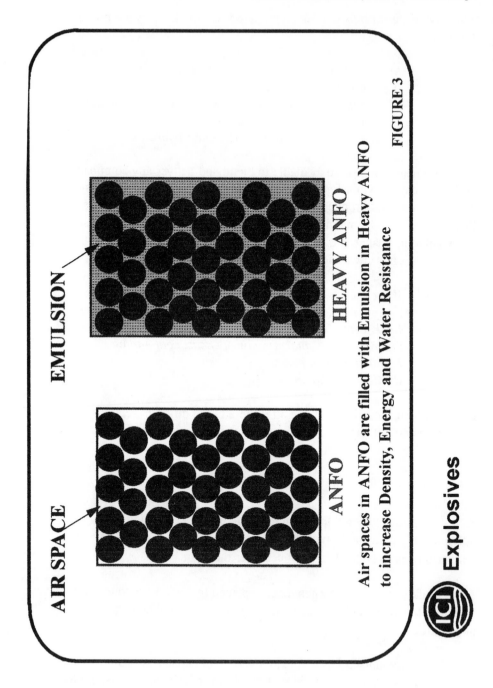

FIGURE 3

Air spaces in ANFO are filled with Emulsion in Heavy ANFO
to increase Density, Energy and Water Resistance

Mixing and loading of shotholes is carried out using push-button and hand-operated controls from a panel. Data loggers are also available to record the quantities of each blend of product used in each hole in a blast, and other information.

Some manufacturers have trailers available for use with emulsion trucks to increase their capacity (figure 4).

Many chassis are available for use on public roads, which at the same time have good off-road capability for site work. This, together with the simple operation of delivery units and their ability to provide a wide range of compositions "at the flick of a switch", has enabled "milk-run" type operations to expand. "Milk-run" operations involve servicing a number of sites, perhaps in a single day, from a single satellite plant. In this way even small users have been able to benefit from bulk emulsions.

A typical emulsion truck with good on- and off-road capability is shown in figure 5. In this case the control panel is positioned at the rear of the unit.

Where site access is difficult emulsion units can be located on an all-terrain chassis such as a CAT D25 articulated chassis (figure 6). Such vehicles tend to be permanently based on a single site where large quantities of explosive are consumed.

6. SMART EMULSION TRUCKS

Smart emulsion trucks incorporate micro-processor or PC/PLC control systems and provide improved quality control of delivered product together with:-

6.1   Programmable Loading of Shotholes

The type and quantity of explosive to be loaded into each shothole is programmed into the truck's control system. The truck then loads the holes automatically in accordance with the programmed loading schedule.

6.2   Data Logging

Data Loggers accurately record the quantity of explosive delivered into each shothole together with the percentage blend of emulsion phase and ammonium nitrate.

EMULSION TRUCK AND TRAILER

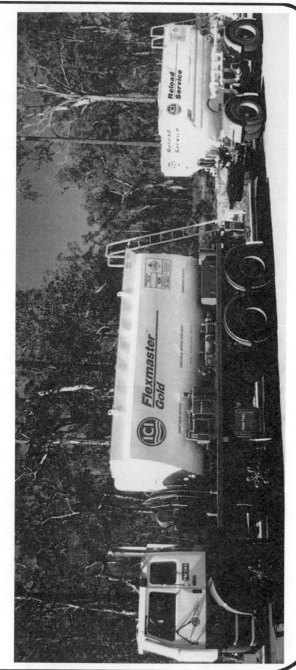

FIGURE 4

 **Explosives**

EMULSION TRUCK FOR "MILK RUN" OPERATIONS

FIGURE 5

Explosives

EMULSION UNIT ON ALL TERRAIN CHASSIS

FIGURE 6

 Explosives

6.3 *The Ability to Interface with Data Loggers on Drills*

This allows information from drills to be downloaded to trucks so that the quantity and type of product loaded into each hole is automatically varied according to depth and the type of strata the hole passes through. In this way more energetic product is automatically loaded adjacent to stronger strata and less energetic product adjacent to weaker strata.

6.4 *The ability to Interface with GPS* Systems on Drills*

This enables the position of individual holes in a blast to be relayed to a truck and then charged in accordance with the loading schedule pre-programmed into the truck control systems or in accordance with information downloaded from data loggers on drills.
*   Global Position using Satellite*

7. UNDERGROUND APPLICATION OF BULK EXPLOSIVES

The application of bulk ANFO in surface operations has been mirrored by its application in underground mines and civil engineering tunnels. The advantages and disadvantages of ANFO in underground mines and civil engineering tunnels are similar to those in surface operations. In most situations the ANFO is either mixed on the surface by the user and transported underground in suitable containers, or it is supplied ready mixed by explosives manufacturers.

ANFO may be delivered into shotholes simply by pouring into down holes or by using pneumatically operated loaders. These loaders can be broadly classified into one of three groups:-

1.   *Venturi*
2.   *Pressure Vessel*
3.   *Pressure Vessel/Venturi*

Figure 7 shows ANFO delivery equipment in use underground.

Unlike surface operations, bulk slurry never achieved widespread use underground, whereas bulk emulsion most definitely has.

ANFO DELIVERY EQUIPMENT IN USE UNDERGROUND

FIGURE 7

 **Explosives**

The advantages of bulk emulsion over bulk ANFO are as follows:-

1. *Significantly reduced concentration of toxic gases in after-detonation fume and when "loading out" (mucking).*
2. *Cleaner operation - less "blowback" and reduced possibility of ground water contamination.*
3. *Improved water resistance.*
4. *Faster loading.*
5. *Easier to control loading, therefore more accurate and precise blast ratios.*
6. *Higher energy product which can reduce the number of shotholes required or provide improved fragmentation/ breakage and thus faster, smoother loading-out with less equipment maintenance.*

Delivery systems vary from a hopper and pump mounted on any suitable carrier which provides electrical supply and hydraulic flow (figure 8), to a custom designed hopper, frame and pump assembly incorporating a hose reeler mounted on a suitable carrier. Hydraulic power may be supplied in different ways, e.g:-

1. *Plumbed into carrier*
2. *Air motor over hydraulic supply*
3. *Independent electric motor over hydraulic supply*

Remote loading arms, either hard wired or remote radio controlled are available. Figure 9 shows a custom designed assembly incorporating a remote loading arm.

As with surface bulk emulsion delivery systems, data loggers are available.

Bulk emulsion phase may be stored underground in tote bins as part of a transportation/storage system used to move large quantities of bulk emulsion from above ground tanker trucks to underground storage areas, safely and efficiently.

Other transportation/storage systems are also available.

EMULSION DELIVERY EQUIPMENT IN USE UNDERGROUND

FIGURE 8

 **Explosives**

CUSTOM DESIGNED ASSEMBLY INCORPORATING REMOTE LOADING ARM

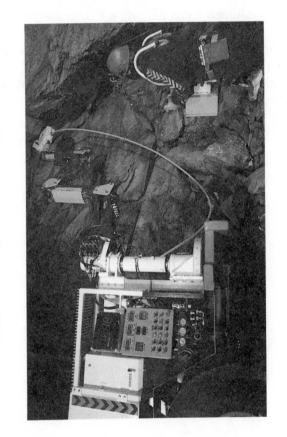

FIGURE 9

Explosives

8. CONCLUSIONS

The use of bulk explosives has expanded rapidly since the introduction of ANFO in the 1950's and the expansion is continuing today with the availability of more and more sophisticated bulk emulsion delivery systems.

Recent trends indicate a replacement of ANFO with bulk emulsions and there is no reason to suppose that this will not continue into the future.

If the 100 years ending in the 1960's belonged in explosive terms to nitro-glycerine, the period since then has most definitely belonged to ANFO and its more energetic and water-resistant derivatives, initially slurry but more importantly emulsion.

If the next 30 years are as eventful as the last 30 years there will be changes indeed!

References:-

(1)   The Science of Industrial Explosives : Cook, Library of Congress Catalogues Card No 74-82419. Printed by Graphic Service and Supply Inc, USA

Session 5. Oilfield Applications and Vibration

Perforating – an Oilfield Application of Explosives

Andrew Pettitt

SCHLUMBERGER WIRELINE & TESTING SERVICES (UK) LTD, HOWEMOSS TERRACE, DYCE, ABERDEEN AB21 0GR, UK

1 Introduction

When one is asked to list the uses of explosives, inadvertently mining, quarrying, weapons and civil engineering spring to mind. The oft forgotten end user of explosives, broad and varied, is the Oil and Gas industry which employs explosives in many varied forms. One of those applications is well perforating. This Paper sets out to summarise how the use and application of shaped charge technology, originally borrowed from military end use, is now an everyday operation used to ensure that we ultimately get petrol in our cars and petroleum products for the vast array of end uses of the petrochemical industry. Once drilled, a well has to be cased or "completed" using a steel liner, this is then cemented in place to save the hole from collapsing. By doing this however, any potential flow path from the oil or gas reservoir to surface is blocked off. The process of perforating re-establishes that path, and permits the hydrocarbon produced to flow to surface.

The perforating process itself has changed significantly over the years as oil field techniques have evolved and requirements have become more demanding. In particular, with wells in the North Sea being drilled to deeper depths, encountering higher bottom hole temperatures and pressures, now as high as 400 degrees F and 20,000 psi, perforating systems containing explosives have to be capable of performing safely and efficiently under these hostile conditions.

The Paper will discuss the particular explosives chosen, and how the very specific requirements of this industry, dictate the types of explosive materials that can be used and how they are tested and evaluated to determine suitability.

1.1 Perforating Techniques

During the evolution of oil well drilling, it became desirable to set metallic casing and cement it through the producing zones, both for well control and production control of oil, gas or water production, and for stimulation, sand control or multiple completion operations. The problem then was to establish communication between the casing and the formation so that productivity was as good as if not better than open hole. Mechanical perforators of various types appeared on the market after 1910 but had limited success. Explosives were in use, but for different uses; they were used in open hole in an attempt to stimulate production by fracturing the formation and increasing hole size. Most commonly used was nitroglycerin. Accidents were predictably common.

The drilling of wells was first documented around 600BCE, when the Chinese were known to drill for gas, although the more lucrative salt market led to a more systematic and technical approach to drilling. By the mid 16th century, wells of 2000' in depth were not uncommon.

Using large wooden rigs, the Chinese perfected percussive drilling using trial and error techniques. A weighted bit was used to strike the bottom of the hole which progressively reduced the strength of the rock until it physically broke up. Percussion drilling or cable tool drilling, as we commonly know it, became the norm in the West, and was used by a Colonel Edwin L. Drake to drill the first recorded oil well at Titusville, Pennsylvania, in 1859. Rotary drilling began around the turn of the century and became successful only after the First World War .

In Drake's time, it was common practice to run a surface pipe to control loose sands and exclude surface water. Beyond this, however, wells were simply drilled into bare rock and left like that. If the bit happened to encounter oil or gas, the fluids would push uncontrollably to surface, more often than not along with a substantial amount of water. It would be up to the ingenuity of the prospectors to cap the well and control production. Drake's well reached a depth of 69 feet on August 27, and since nothing noticeable occurred, the well was abandoned. Days later it was noticed that the well had filled with oil and production ensued.

The dramatic pictures known to us all where oil comes gushing from the well during drilling operations arises when the drill bit strikes reservoir pressure, dramatically releasing an uncontrolled flow of oil to surface.

Various techniques were used in the early days to control and isolate the well pressure, but it was as the drilling became more technical, running steel liners or casing through a producing zone became standard practice, particularly in areas of unconsolidated sands. By 1919 all sorts of screened pipes were available for the purpose, but blank pipe was also used. Mechanical perforators that cut holes through steel became available sometime after 1910, but had limited success. As late as 1930 there were numerous experiments with mechanical casing perforators, albeit unsuccessful.

Not all the early oil wells were gushers, and while much effort was channeled into controlling the blowouts, a parallel group of people were developing techniques to enhance the wells which failed to produce. In 1866, a Col. E.A.L. Roberts perfected a method of stimulating poor producers by detonating gun powder packed into torpedoes that were lowered opposite the formation. Using a simple drop bar to detonate his charges, Roberts convinced a sceptical audience by stimulating a dry well to produce 80 barrels a day with two separate explosions. Since drop bars have a tendency to get stuck in the hole, nitroglycerin time bombs were developed and considered more reliable. They were used as late as 1928 in Oklahoma and Texas. Brand names of these devices included "Zero-Hour Electric Bomb", "The Bolshevik", and "The King". These devices are known to have been used as late as 1951.

For the record, acidization was first used for stimulating wells in 1895, but then promptly forgotten. The idea was resurrected in the late 1920s by the Pure Oil Company and Dow Chemical, which had surplus hydrochloric acid on hand. By 1932, acidization had earned enough commercial success for Dow Chemical to form the first acidizing company, Dowell Inc. Fracturing began after the Second World War and was initially commercialised by the Halliburton Oil Well Cementing Company.

As the technique and popularity for running the steel liners grew, so did the quest for developing efficient techniques for punching holes through all the steel and cement to re-establish flow of the hydrocarbons. Figure 1 shows the typical wellbore geometry of a perforated well.

Mechanical devices were used with mixed success, and it was the bullet gun that was adopted as the next in vogue technique. Sid Mims, a Los Angeles oil man, claimed a patent in 1926, but it wasn't until 1932 that a device was actually built and tested. Detonation was achieved by passing an electric current down a conducting wire. It took 11 runs to fire 80 shots, and the well, once written off as uneconomical, began to produce commercial quantities of oil. All this was the effort of two men who had acquired Mims's patent, Walter T. Wells and Wilford E. Lane. Their company, Lane-Wells, acquired many years later by Dresser Industries, Inc. was the first to offer perforating services to the oil industry. Figure 3 illustrates an example of a Bullet Gun.

Bullet perforating was still popular into the 1960s, and to this day bullet guns remain a service of some perforating companies for special applications. The heyday of bullet perforating was undoubtedly in the early 1950s, just before shaped-charge technology took the ascendancy.   In soft formations, these guns were particularly effective, but in hard rock the bullets might not penetrate even the casing, and in such instances it was not uncommon to bail bullets out of the well. Worse yet, they could jam the gun.

With the advent of the Second World War, technological innovation and investment in the use of explosives for military applications had progressed dramatically. With the development of the shaped charge, and its ideal performance characteristics, the perfect tool for creating the necessary holes in oil wells had been discovered. How to harness and fine tune this technology was to be the focus of great effort over the next 50 years.

There are two generic methods of perforating wells today using shaped charge technology, namely Wireline Conveyed Perforating (WCP) and Tubing Conveyed Perforating (TCP).   As designated by the name, the difference lies in the method of conveyance of the perforating guns in the well. For WCP, an electrical cable is used, but with TCP, the guns are an integral part of the testing or completion string. Supporters of each technique will compare the merits and drawbacks of their own preferred method, a few are noted below :

1.2   Wireline Conveyed Perforating

1.2.1 Advantages
- Quicker trips in and out of the well
- Explosives exposure downhole limited
- Simple gun handling procedures
- Logistically lighter operation

1.2.2 Disadvantages
- Due to weight restrictions, limitations exist for lengths of guns per run
- WCP guns tend to be smaller, therefore performance limited due to charges used
- Debris left in wells can be higher from strip type guns
- If long intervals have to be perforated, operations are timely and costly due to multiple runs
- Perforating at high pressure with wire in a well can create problems

1.3   Tubing Conveyed Perforating

1.3.1 Advantages
- Guns used are at a higher shot density and give greater performance
- No limit on length of interval that can be perforated in a single trip
- Technically more adaptable for deviated wells
- Short intervals can be less cost effective with TCP
- Radios and other electrical signals are not a safety hazard
- Positive confirmation of detonation often difficult
- Better well clean ups possible resulting in optimal performance

1.3.2 Disadvantages
- Operations will generally take longer with TCP and this can have cost implications if rig costs are high
- Temperature requirements often more complex
- Long recovery times for any misfires
- Higher personnel requirements
- Short intervals can be less cost effective with TCP
- Positive confirmation of detonation often difficult

Figures 4 and 5 show the different types of guns that are commonly available.

1.4   Explosive Selection

A key point in selecting explosives for well perforating is the presence of the elevated temperatures that are seen downhole. Being chemical compounds, as exposure to temperature exceeds a certain level and duration, a degradation and reduction in stability is seen. With safety, reliability and effectiveness being the prime considerations, a suitable explosive has to be selected such that temperatures can be tolerated, often for prolonged periods of time. Figure 6 shows a summary of the secondary high explosives that are potentially suitable for oilfield use. The table reflects the performance of explosives after various temperature/time exposures. These data are acquired by subjecting   charges to a variety of temperatures for the range of times shown, and evaluating the performance and stability. The published performance tables for each vendor, will reflect the types of test done, the purity of the explosives chosen and the safety factor adopted.

In today's industry, the commonly used explosives are RDX, HMX, PYX and HNS.

RDX = Cyclonite
HMX = Octogene
HNS = Hexanitrostilbene
PYX = 2,6-Bis(Picrylamino)-3, 5-Dinitropyridine

Selection of the ultimate charge and explosive type is generally based on commercial and proprietary reasoning.

1.5   Charge Performance

There are two basic requirements from oil field shaped charges for perforating, namely, either deep penetration, or, for special applications, a resultant big hole.

For the end user of perforating charges, the accepted method of charge comparison is the API RP 43 standard which is now in its Fifth Edition. This standard was formulated by the American Petroleum Institute and lays down the parameters by which all explosive charges can be metered against others. There are several tests done, but the main one employed today is the Section 1 test which involves firing a charge into a concrete target which has been suitably cured, and which exhibits a particular compressive strength. Figure 7 shows the test set up.   Each vendor then presents his overall data for due consideration by the end user. Engineers requiring charges for particular applications can then access these data and select the charge and appropriate perforating gun accordingly. One limitation of the API tests is that the performance of each charge is taken at ambient conditions on surface, whereas end use is downhole under harsh conditions. In order to alleviate this partially, there are a number of proprietary software packages that allow some degree of downhole performance prediction. These softwares use various processes including algorithms and test works which are combined to give a simulation of actual charge performance.

1.6   Methods of System Detonation

Early techniques for the detonation of perforating guns relied solely on electrical methods, dictated solely by the electric wireline method of conveyance.   The major drawback for electrical detonation on an oil installation can be the considerable disruption created by the need for radio silence to protect operations from stray currents. This affects operations such as helicopter flights, standby boats, welding work, cathodic protection devices, radio communications and other associated hazards. Not only does this shut down cause some degree of time delay, but there are cost implications. With mobile oil rigs costing anything up to £100,000 per day to operate, and some perforating operations taking several days, hidden costs are significant. This drawback has now been solved, with the adoption of the "slapper" type detonators which utilise a foil initiation system. Many perforating operations requiring wireline conveyance now use this method.

For the tubing conveyed technique of perforating there are several techniques that have been used over the years, namely :

- electrical methods using conventional "wet connect" type heads;
- pressure methods   direct pressure fired heads;
                                        differential pressure fired;
- mechanically fired   drop bar methods;

The first systems introduced by GeoVann (now Halliburton), and originally the only method available for starting the detonation train was a mechanical means, and this involved dropping a steel bar from the rig floor, down on to the percussion firing cap. Figure 8 shows an example of the GeoVann Mechanical Head. Bearing in mind that there was often in excess of 15000 feet between the rig floor, along with high deviation, perhaps irregularly profiled surfaces on other tools, it was not surprising that sometimes there were difficulties associated with drop bar techniques. GeoVann then embarked on a notable quest to perfect the art of gun detonation methods and to this day take most if not all of the credit for introducing alternatives to the bar drop. First advancements to the bar drop was a direct pressure firing head which functioned after surface applied pressure sheared out steel retaining pins holding a piston proud of an initiator. This proved to be a very acceptable solution and was followed by a differential firing technique which again offered more flexibility for the Operators. After some years of greatly improved reliability, more companies entered the perforating market bringing with them new firing systems. One notable step forward was introduced by Dresser Compac who developed a retrievable family of firing heads (Figure 9). The advantage of this was evident. Perforating guns could now be run into a well, minus the primary high explosive firing head, and located only when required. Additionally, if there were any problems in the firing operation, the head could be retrieved from the well, inspected and if necessary, replaced.

Further landmarks in firing head technology include the introduction of dual firing heads which allow more than one type of firing head to be deployed at one time. Figure 10 shows the Schlumberger Dual Firing Head layout. Probably the latest technology to be used is a pressure pulse telemetry method. Here a pressure pulse or fingerprint, is applied to the wellbore. The firing system is programmed to recognise various pressure pulses at specific amplitudes and durations, which are applied at specific temperatures and times, before it is armed. The guns are then fired using an additional pulse. Reasons for this being introduced include immunity from the many other pressure operated tools which are now included in downhole operations, and also the difficulties in using conventional pressure fired heads in horizontal wells, due to access.

1.7 The Future

With the higher temperatures encountered in recent years, statistical probability of system failure has risen. Also when failures of perforating guns occur in these environments, they are normally very notable and extremely expensive to rectify. In an attempt to reduce the frequency of failures, an industry group was set up by a group of Operating companies predominantly involved in the high profile HPHT(high pressure, high temperature) sector. This is known as the PEGS (Program to Evaluate Gun Systems) Group. Its prime function is to define standards and implement procedures for verifying that perforating systems will perform to temperature, pressure and time specifications on an ongoing basis. The areas under investigation and review by the PEGS team include :

1.7.1 Explosive Test
- Production Line QC Target
- Production Line Test
- Set Up Production Line Charge Performance
- Charge Performance
- Thermal Integrity Test
- Material Traceability
- Packaging Control

1.7.2 Gun Hardware
- Pressure / Temperature Test
- Dimensional Control
- Mechanical Control (Chemical and Physical Properties)

- Thermal Integrity
- Mechanical Integrity
- Elastomer Control

1.7.3 Accessories

- Pressure / Temperature Test
- Verification of Functioning
- Assembly Inspection

The above program is a continually evolving process but highlights the importance placed on perforating and explosives systems that are now being used to further the production of oil and gas wells.

Time will undoubtedly see the frontiers of temperature stability pushed beyond the current limitations, and it is no longer a question of purely explosive reliability, but also the many other components that make up the perforating system, and upon which we place great faith in reliability and performance.

Conclusions

As the oil and gas industry becomes more mature, bringing with it the need for more complex technical solutions, the conditions that we expose our explosives to are becoming far more severe, and closer to the edges of known technology. In the search for oil, wells are now being drilled deeper, with the knock on effect that the downhole temperatures and pressures that will be experienced are much higher, pushing them beyond what we have previously experienced. The next phase of reservoir developments sees the need for our perforating charges, and the explosive chosen therein, to be stable for longer periods of time, at higher pressure and temperature, yet at the same time being able to perform as well as, if not better than, current products. This requirement in itself will see the industry strive to engineer solutions unique in the field of explosives.

What specifically does the future hold for perforating? Probably the main requirement will be for suitable explosives that can be used at temperatures in excess of 500 degrees F for a period of 100 to 200 hours, dependent on operational objectives and specific requirements. In being able to provide this, with the suitable and stringent QA/QC program to match, the industry could rightly claim to be pushing the frontiers beyond what has been conventionally expected from it. Only time will tell how long it will take to be reality.

Acknowledgments

The author would like to acknowledge with grateful thanks the Management of Schlumberger Wireline & Testing for allowing permission to present this Paper, with particular mention to Jack Lands, Jorge Lopez de Cardenas of the Schlumberger Perforating & Testing Centre, Rosharon, Texas, USA. Thanks are also due to Terry Digges of Halliburton, Aberdeen, and Roger Pelly of Western Atlas, London, who have both assisted in the preparation of this Paper by providing information on Firing Head technology.

References

1. WP Walters & JA Zukas, Fundamentals of Shaped Charges - ISBN 0-471-62172-2
2. Rudolf Meyer, Explosives (Third Edition) - ISBN 0-89573-600-4
3. S Fordham, High Explosives and Propellants - ISBN 0-08-023833-5
4. Services Textbook of Explosives - JSP 333
5. Perforating Services Manual - Schlumberger Publication SMP-7043

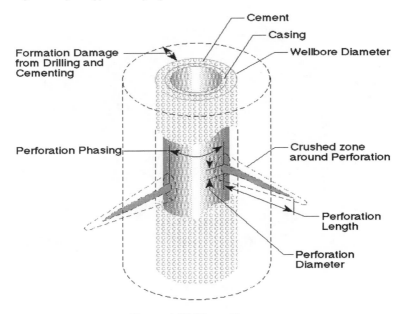

Figure 1 *Wellbore Geometry*

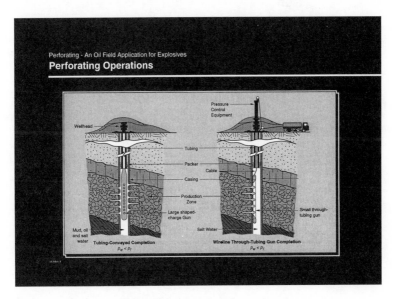

Figure 2 *Perforating Operations*

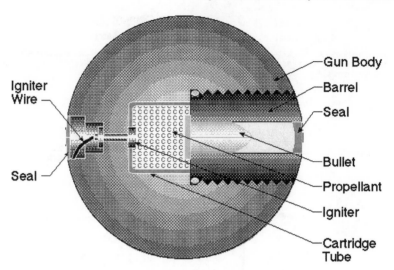

Figure 3 *Bullet Gun*

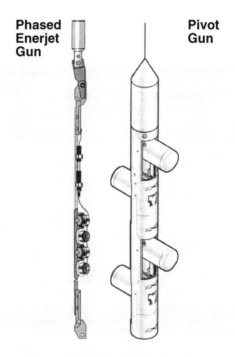

Figure 4 *Wireline Perforating Guns*

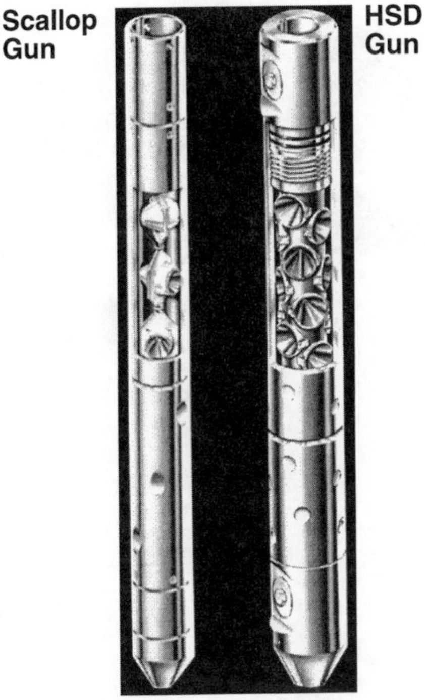

Scallop Gun **HSD Gun**

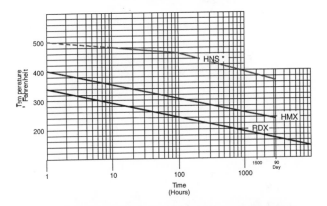

Figure 5 *Carrier Guns*

Explosive Type	1-hr rating	100-hr rating
RDX	340°F [166°C]	240°F [115°C]
HMX	400°F [204°C]	300°F [149°C]
HNS	500°F [260°C]	460°F [238°C]

Figure 6 *Temperature / Time Chart for Oilfield Explosives*

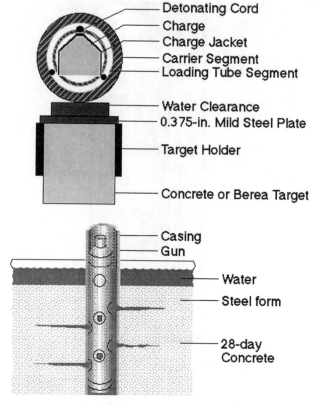

Figure 7 *Concrete Target and QC Test Configuration*

Figure 8 *GeoVann Mechanical Firing Head*

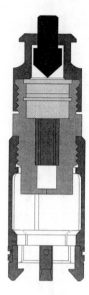

Figure 9 *Dresser Compac Retrievable Firing Head*

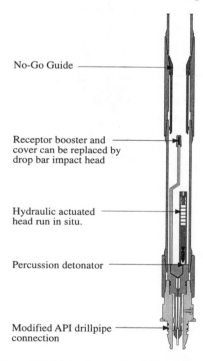

No-Go Guide

Receptor booster and
cover can be replaced by
drop bar impact head

Hydraulic actuated
head run in situ.

Percussion detonator

Modified API drillpipe
connection

Figure 10 *Schlumberger Dual Firing Head*

Environmental Effects of Blasting – Recent Experiences

R. A. Farnfield

DEPARTMENT OF MINING AND MINERAL ENGINEERING, UNIVERSITY OF LEEDS, LEEDS
LS2 9JT, UK

1 INTRODUCTION

The Department of Mining and Mineral Engineering at the University of Leeds has been
carrying out research into the monitoring, prediction and control of the environmental
impact of blasting for nearly 20 years. During that time many topics have been covered
including :

- Improved Prediction Techniques.
- Improved Control Technology.
- Development of Monitoring Equipment.
- Structural Response Modelling Using Response Spectra.
- Structural Response Modelling Using Transfer Functions.
- Structural Damage Criteria.
- Human Response Monitoring.

It is widely recognised that there are two problems associated with the environmental
impact of blasting. Most research work has concentrated on the possibility of structural
damage but it is well known that the overall problem is one of human response. The
University of Leeds has a continuing programme of research in these two fields and it is
this work which will be described in this paper.

2 BLAST VIBRATION STRUCTURAL DAMAGE RESEARCH

The University was funded by British Coal Opencast to investigate the effects of surface
blasting operations on a domestic building. Although similar work has been carried out
elsewhere, as summarised by Siskind, Stagg, Kopp and Dowding[1], this was the first study
of its type in the U.K. The aim was to identify all the cracking which took place in two
rooms during the 30 months of the project, and to investigate the possible causes.

The work was carried out in two phases:

i)   With the blasting and operations distant from the property (less than 2mm/s resultant PPV at the foundation) and crack surveys carried out every four weeks.

ii)   With operations approaching the property (PPVs exceeding 2mm/s for the first time) and pre- and post-blast crack surveys for all blasts.

This paper gives only a summary of this work, full details of the monitoring procedures and equipment are given by Farnfield and White[2] with detailed information on blast induced damage by White, Farnfield and Kelly[3].

2.1 Experimental Details

The project took place on a production opencast coal site at the western end of the South Wales Coalfield in the United Kingdom. The property was situated in the middle of the site and was due to be demolished during the operation of the mine. An experimental procedure and monitoring equipment were designed to enable damage to be identified and recorded and the environmental and structural influences to be assessed.

2.1.1 Site Description. The geology of the site was highly complex with steeply dipping seams which were intensely folded and faulted. The blasting took place on various horizons with benches which were usually 6m high. A mixture of electric and non-electric detonators was used and ANFO was the bulk explosive. Blasting operations passed within 200m of the building during Phase 1 before moving further away. During phase 2 of the project blasting took place close to the property over a period of several months.

2.1.2 Property. The house was a stone built cottage with a two storey concrete block extension. A picture of the property is given as figure 1 and shows the house as it was towards the end of phase 2 of the project.

Prior to taking possession of the structure its roof had been removed leaving the majority of the building in poor condition. The condition of the cottage meant that the study could only take place in two rooms of the extension; one upstairs and one downstairs. The variety of wall types in the two rooms included plastered double-skin concrete block, plastered single-thickness internal wall, the old stone wall of the cottage which had been smoothed with concrete and then plastered, and finally some partition walls with plasterboard and a plaster skim. All wallpaper was removed so that the plaster itself could be inspected.

2.1.3 Surveying. During Phase 1 a full crack survey took place every four weeks. All new and extended cracks were marked and recorded using a combination of visual inspection, photography and computer digitisation. As the blasting got closer in Phase 2 the vibration levels rose and pre- and post-blast surveys were carried out for every blast. Figures 2 and 3 show the crack plots for a wall in the upstairs test room from the beginning and end of the research project. The additional cracking shown has been proven to be due to the fan heaters in the house, timed to come on twice a day in the winter, one of which was pointing at this wall.

Figure 1 : Test House in the Middle of a Surface Coal Mine, Photographed in Phase 2 of the Research Project.

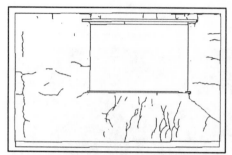

Figure 2 : Upstairs Bedroom Wall Crack   Figure 3 : Upstairs Bedroom Wall Crack
Plot at Beginning of Project.   Plot at End of Project.

2.1.4 Vibration Recording. Four sets of triaxial transducers were used to continuously monitor the house along with an air overpressure transducer. Two sets of velocity transducers and two sets of accelerometers recorded the vibration at various points around the house and these were linked to digital blasting seismographs. A typical set of vibration recordings is given in figures 4 and 5 showing relatively high vibration levels on structural members compared to the foundation level.

2.1.5 Structural Data. Changes in 11 crack widths were monitored using linear variable displacement transducers and a record taken every 20 minutes. A measurement was also taken after each vibration event. To provide continuous data on house movement, two precision inclinometers were installed on the ground floor. Precision level surveys which monitored ground and house level took place on a 4-week basis in Phase 1, but then were carried out weekly in Phase 2.

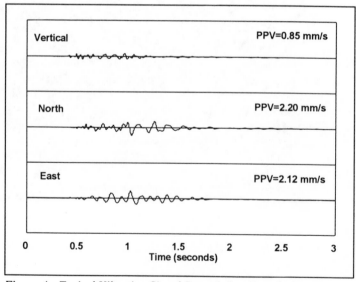

Figure 4 : Typical Vibration Signal Recorded at Foundation Level.

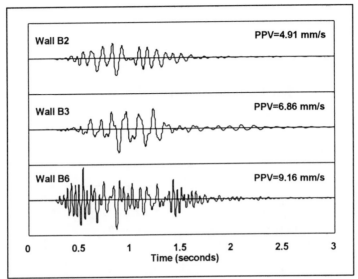

Figure 5 : Mid-Wall Response to Vibration Shown in Figure 4.
Wall B2 is an External Concrete Block Cavity Wall.
Wall B3 is an Internal Rubble Wall.
Wall B6 is an Internal Stud Partition Wall Covered in Plasterboard.

2.1.6 Environmental Data. A full weather station recorded the maximum and average wind speed and direction for 20 minute periods along with the external and internal temperatures and the rainfall. An attempt was made to monitor humidity using an aspirated psychrometer, but severe problems were encountered with the wick supplying the wet bulb continually drying out. Piezometric water level was monitored every 20 minutes in a borehole immediately adjacent to the property.

2.2 Phase 1 Results

The 24 months of monitoring in Phase 1 was used to determine the influences other than blasting which could cause damage. During this period the resultant PPV in the ground next to the property never exceeded 2mm/s. Although there was no major structural damage there was cosmetic plaster cracking. Extensive analysis was carried out to determine which factors influenced the cracking rate.

Figure 6 shows a graph of crack width and internal temperature over a four day period during which time the heating system was set to switch on twice a day. It can be seen that the crack widths are cyclical in nature and inversely proportional to internal air temperature which means the cracks are widest and the stresses greatest when the temperature is lowest.

Figure 7 shows a graph of crack increase in the two test rooms against time, crack increase is shown to be greatest during the winter months and especially when associated with an operational heating system. The exact mechanism for the increase in cosmetic cracks was not determined but was thought to be a combination of very cold temperatures causing stress with periods of heat drying the plaster out and increasing its brittleness.

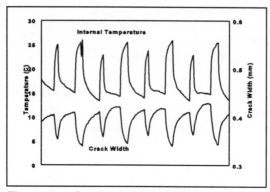

Figure 6 : Graph of Crack Width and Internal Temperature Over A 4 Day Period, Showing Temperature Dependency of Crack Width.

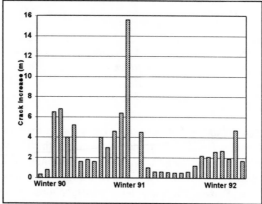

Figure 7 : Total Crack Length Increase Per 4 Week Survey for Both Test Rooms Covering 3 Winters.

2.3 Phase 2 Data

Phase 2 of the project commenced as the blasting started to approach the property. PPV levels gradually increased to exceed 5mm/s for the first time (8.68mm/s) and pre-and post-blast surveys commenced.

On 2/4/92 a blast took place very close to the property and resulted in a resultant PPV of 24.1mm/s. This was the first blast over 10mm/s and another one later that day was the largest that the house experienced at 59.87mm/s. From then until the house was demolished on 10/6/92 there were a total of 78 blasts, 12 of which were over 10mm/s.

Table 1 shows the 12 events over 10 mm/s PPV with crack length data. Pre-blast damage is that increase in crack length surveyed prior to the blast event and gives an indication of how active cracking was at the time of the blast. All damage noted here is cosmetic in nature and would not have normally been noticed under a wallpaper covering.

It is apparent that there is no simple relationship between PPV values and damage as there is cracking found on the studied walls after a blast of 14.3 and 17.3mm/s but not after a 33.5 or 34.7mm/s, although these blasts did damage elsewhere in the house due to failure of the pillar upon which the house was then perched. After these 2 blast events, shown by a double line in table 1, the structure was considered to be very unstable due to pillar failure and particularly sensitive to vibration damage. Even without taking into consideration the sensitive condition of the structure it can be seen that the lowest PPV which caused an increase in cosmetic crack length was 14.3 mm/s.

It is interesting to note that most of the damage which was evident in the property occurred on walls in the downstairs room which is likely to have been subject to lower absolute vibration levels than upstairs in the house.

During Phase 2 of the project total crack increase in the structure was 15.55m with only 3.45m, or 22%, being possibly attributable to blast vibration.

Date	PPV mm/sec	Pre-Blast Damage mm	Post-Blast Damage mm	Damage Location
2/4/92	24.1	----	62	A1
2/4/92	59.9	----	268	A3,B2
25/4/92	11.7	----	----	No Damage Caused
27/4/92	17.3	410	724	A1,A4
28/4/92	21.7	946	777	A1,A2,A4,B1,B3
29/4/92	18.7	----	----	No Damage Caused
1/5/92	12.5	----	----	No Damage Caused
5/5/92	34.7	----	----	External Damage
7/5/92	33.5	----	----	External Damage
8/5/92	26.1	----	1137	AB,B2
19/5/92	14.34	----	486	A3,B1,B3
9/6/92	37.4	----	----	No Damage Caused

Table 1   List of vibration events over 10 mm/s PPV as measured at foundation level with crack length information and damage assessment.
Walls A1,A2,A3,A4 and AB are in the Ground Floor Room.
Walls B1,B2 and B3 are in the First Floor Room.

2.4 Structural Damage Research Conclusions

It has been demonstrated that a typical domestic property is subject to increased levels of cosmetic plaster cracking as a normal part of its existence and that such crack increases can be closely associated with climatic conditions and in particular temperature.

The lowest resultant foundation PPV value after which damage was found was 14.3mm/s although the house was shown to be very unstable at this time. The lowest PPV blast which definitely gave rise to damage was 24.1mm/s, even though the condition of the house was fairly poor. All these values are greater than are normally permitted from blasting operations in the U.K.

3 THE ACOUSTIC RESPONSE OF STRUCTURES TO BLASTING

Given the findings of the research outlined in section 2 of this paper, and much other evidence [1], it is clear that complaints from members of the public in the U.K. concerning blast induced vibration are not the result of actual structural damage but rather due to adverse human response and the fear of structural damage.

It is also a well known fact that the majority of complaints from members of the public are as a result of that person experiencing the response of a structure to a blast. This leads to a question of exactly what a person experiences inside a building subject to the blast induced ground vibration and air overpressure.

In general the environmental impact of blasting is considered to be split into two aspects, air overpressure and ground vibration. The response of structures to these two different types of vibration is relatively well documented and understood [4,5]. Many complainants mention 'rattles' and other such structural noise and this is most often assigned as being a side effect of air overpressure.

The University is currently undertaking research into such structural noise or acoustic response in order to prove that it exists, determine what causes it and investigate the factors by which it may be controlled. To date this work has been centred on a monitoring programme in a house close to an operating surface coal mine in the U.K.

3.1 Experimental Procedures and Set-up

3.1.1 Test Site Description. The test structure consists of a two storey domestic building in the form of a terrace of small houses. The room used for recording was on the 1st floor, had a wooden suspended floor, a single external wall and one window. The room was unfurnished and without a carpet. Typical background noise levels inside this room were found to be between 35 and 38 dB(A).

3.1.2 Monitoring Equipment. The test house was instrumented with computer based blast monitoring equipment. This included a tri-axial geophone unit at foundation level, a similar unit in the middle of the upstairs test room, an external air overpressure transducer and an integrating sound level meter also in the upstairs test room.

The equipment recordings were triggered by ground vibration and recorded all parameters to disk including the vibration signals and the 'A' weighted sound level reading from the sound level meter. During the monitoring period in the region of 150 blast events were recorded with this equipment.

3.2 Experimental Results and Analysis

3.2.1 Example Recordings. Figure 8 shows a typical recording from this test location. The vibration data displayed is from the transducer mounted in the room itself. Peak values in this example are 8 mm/sec resultant and 28 Pa (123dB(Linear)) for air overpressure. The maximum induced noise is 67 dB(A). It can be seen that the rise in noise levels coincides with the arrival of the vibration and that the air overpressure signal has little effect on the noise level and this was found to be typical of all the recordings made.

What is immediately obvious about this recording is the level of noise induced within the room with 60dB being exceeded for approximately 1 second. Penn[6] gives a table of example noises with 60 to 70dB being typical of the levels for normal conversation. Given

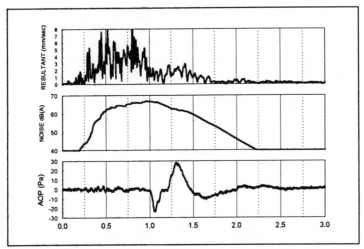

Figure 8 : Recording From First Floor Room Showing the Mid-floor Resultant Vibration, Acoustic Response and External Air Overpressure.

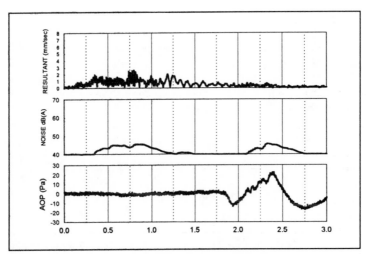

Figure 9 : Recording From First Floor Room Showing Low Level Response to Both Vibration and Air Overpressure.

a background level of less than 40dB(A) this blast caused an increase in noise of more than 27dB in this room.

Figure 9 shows a recording with a relatively low resultant PPV of 2.5 mm/sec but with a relatively high peak overpressure reading of 22Pa (122dB(Linear)). The acoustic response peaks at 44dB(A) for both the vibration and air overpressure. This recording illustrates that the acoustic response can be due to both vibration and air overpressure.

3.2.2 The Relationship Between Noise and Vibration. Figure 10 shows a plot of the relationship between resultant PPV and induced noise levels for the 150 blast events recorded in the test room. The upper limit line of 71dB(A) is due to the limited range of the A/D convertor board used in the monitoring computer. Data points at this level indicate an overload of the recording system. This figure shows a basic correlation between PPV and noise but the nature of the relationship is not clear. There appears to be a curved lower boundary line below which no data points exist but above this line the data is highly scattered.

It must be emphasised that this room is a special case, being completely unfurnished, and is likely to have produced a higher acoustic response than a typical room. It should also be noted that the vibration levels recorded are taken from the middle of the wooden floor and that the corresponding levels at the foundation are likely to have been much lower.

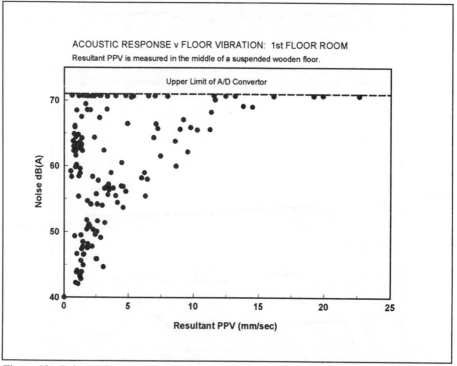

Figure 10 : Relationship between Mid-floor Peak Resultant Vibration and Acoustic Response.

3.3 Conclusions on Acoustic Response Research

With the data recorded in this research project there is no doubt that structures can respond acoustically to blast induced ground vibration. The data also suggest that such a response can be highly noticeable with typical levels being equivalent to normal conversation.

The author believes that acoustic response could be the missing link in our understanding of the reasons why people complain about blasting. The research is continuing to investigate this phenomenon with further field recordings being made at a different location and with complete digital noise recordings being made.

4 CONCLUSIONS

The University of Leeds Department of Mining and Mineral Engineering has a long history of research into the environmental impact of blasting. Significant work has been completed concerning the level of vibration likely to cause structural damage. The problem of human response is a topic of continuing research albeit with no funding.

5 ACKNOWLEDGEMENTS

The type of research work described in this paper does not come cheaply and the author would like to express his gratitude to the former British Coal Opencast for its support over a number of years in the work on structural damage and the continuing support in kind by RJB Mining and Celtic Energy connected with the acoustic response research.

The opinions expressed are those of the author and are not necessarily those of the University of Leeds or any co-operating organisations.

References

1   D E Siskind, M S Stagg, J W Kopp and C H Dowding. 'RI8507 Structure Response and Damage Produced by Ground Vibration from Surface Mine Blasting'. United States Bureau of Mines, 1980

2   R A Farnfield and T J White. 'Research Into the Effects of Surface Mine Blasting on Buildings : Long Term Monitoring Projects'. *The Mining Engineer*, 1993

3   T J White, R A Farnfield and M Kelly. 'The Effects of Surface Mine Blasting on Buildings'. Proceedings of the Fourth International Symposium on Rock Fragmentation by Blasting, Vienna, July 1993. pg 105 to 111. Pub Balkema.

4   D E Siskind, V J Stachura, M S Stagg, and J W Kopp. 'RI8485 Structure Response and Damage Produced by Airblast from Surface Mining'. United States Bureau of Mines, 1980

5   R A Farnfield. 'The Application of Transfer Functions in the Prediction of Structural Response to Blast Induced Ground Vibration'. Proceedings of the 10th Annual Symposium on Explosives and Blasting Research, International Society of Explosives Engineers, Austin, Texas, 1994

6   C N Penn. 'Noise Control : The Law and Its Enforcement'. Pub Shaw & Sons, ISBN 0 7219 0831 4

Session 6. Blast Performance

Experiences with Expert Explosive Electronic Detonators

Vivian Patz

EXPERT EXPLOSIVES, POB 10, MODDERFONTAINE 1645, SOUTH AFRICA

The centenary Commemoration of Alfred Nobel's death is an occasion at which to relate the story of how a simple idea has been refined and developed into a revolutionary computer controlled rock breaking system.

This COMPUTER AIDED BLASTING (CAB) system programs the firing time of a small electronic device (chip) in each detonator. By changing the drawing of a blast on the computer screen, the mining engineer can vary the time relationships between blast holes in the rock.

Driven by the need to make full use of the newly installed South African semiconductor lines, and armed with the idea of initiating explosives directly off the surface of a silicon chip, a bunch of electronic engineers and physicists joined up with AECI to form Expert Explosives in 1988. The enthusiastic team attacked the problem, oblivious to many of the earlier failed attempts at electronic detonators.

The central patented idea was a small low energy integrated structure to fire explosives in intimate contact with the micro-chip, optimising production cost and energy storage requirements through the use of integration.

Progress was rapid and the chip was developed in parallel with the explosive interface, harness connectors and control systems.

In 1991 the first explosive trials showed that sympathetic detonation occurred between adjacent charges. It emerged that the triggering mechanism was the coupling of the electrical energy generated by the first detonation through the wiring harness into the second detonator. This EMP, short for electro-magnetic pulse (a military term used to describe the destruction of electronic circuits by the voltages associated with a nuclear explosions), became the target of our efforts.

The solution required careful measurement of the phenomena   and extensive re-engineering of the electronics, but by the end of 1992 hundreds of units were being successfully fired on the ranges in Modderfontein. Range testing showed up many areas of system integration that needed improvement, especially the wiring and connection systems and the mechanical construction of the electronic assemblies. Nothing could dampen our spirits though - the system concept had been demonstrated. It was possible to use complex electronic devices to accurately control blast timing, and the delays could be generated remotely and sent to the detonators just prior to blasting.

The system was first used in 1993, then qualified for use in quarries, and later in overburden stripping in large open cast coal mines in the Witbank coal fields. Blast size increased rapidly with typical dimensions being 350 holes, each 10 meters apart and 30 meters deep. Wire lengths for the whole blast began to exceed ten kilometres, and the number of connections required the skill of a telephone linesman to ensure success during hook up. The carefully engineered electronic interlocks on the chip appeared to work well, and no safety incidents emerged throughout 1993. The interlocks worked on multiple redundancy.

The on chip bridge could only be fired by the energy stored in a capacitor. The capacitor was shorted by a switch under remote control. Another switch allowed energy into the capacitor. Thus to fire a device, the three switches would have to operate to un-short the capacitor, charge it up and discharge it into the bridge. Coded signals were sent to sequence the switches.

We all learned a sobering lesson on the morning of 31 December 1993. The control box , originally used to test detonator operation only from a place of safety, had been moved onto the bench as hook-ups became more complex, and the confidence of the engineers grew in the safety of their own interlocks. Disaster struck when a second row hole detonated during testing while one of our engineers was standing 2 meters from the blast hole. The hole contained 1000 kg of explosives, was 30 m deep on a 10 m burden and spacing. The engineer flew into the air and landed on a pile of drill chippings, while the blast broke through the front row and blew the high wall across the pit. No cratering around the top of the hole was evident, and miraculously the engineer only suffered a cracked ankle bone and no other injuries. An intensive investigation of the incident indicated that mechanical forces probably resulted in a wire shorting internally in the tube to bypass the interlocks, and it became clear that interlocks were not enough to protect life under all failure conditions.

The concept of inherent safety was invoked by ExEx and made public as the recommended way to safeguard personnel during the development and use of electronic detonator systems. The detonators may not be connected to any circuit which has a voltage higher than the test voltage while personnel are present. The test voltage is unable to fire the bridge, even if all protective interlocks are not operational. This test voltage is inherently safe. Simple. Effective.

By late 1994 we were back on the mines and the system was establishing a good reputation for precision control of timing in difficult blasting conditions. Several blasts were carried out to save crumbling pre-drilled ground in large open pit diamond mines. These massive blasts, consisting of multiple rings, had up to 450 holes. Holes were fired as close as 2 ms. apart, to keep the vibration frequency away from the pit resonance frequency, saving the mine massive ore losses and secondary breaking costs.

Work in open cast mines and quarries was underway to determine how timing should be controlled to yield benefits to the mine. Several interesting factors emerged that obscured our ability to adequately monitor the results of blasts. Instrumentation had to be assembled from several other areas, and multiple types had to be used simultaneously to gather sufficient information about blast effects. Poor priming, charging and stemming practices resulted in several misfired holes in the early blasts.

Measuring downstream effects like truck loading times and dragline efficiencies indicated operator variations that masked blast effects. With the CAB system, it was easy to blast the holes in any order, and typical attempts ended up with delay patterns that copied the timing of the present shock tube initiation systems. The results matched the results of the present systems.

People were reluctant to change timing, since they had no feel for what the changes would do. ExEx responded by working with ICI master blasters to design parameter driven blast design capability into the graphical blast design software - Winblast.

Instead of thinking of delays between holes, the blaster could now think in terms of time contours and throw vectors. The system is still quite primitive, but will grow as more rules are built in. System users prefer parameter driven blast design, and this has led to some very imaginative blast effects.

One such effect was when an operator inadvertently reversed the blast direction of the back row of holes in a colliery blast. The blast sequenced exactly in reverse order. This user error resulted in the most awesome pile of building sized boulders one could ever imagine. Repairing the damage was not easy and it made a major negative contribution to our early operating profits.

The existing system has now been used in all major areas of blasting, and results show that vibration, fragmentation, throw and wall damage can all be controlled to some degree by controlling the timing of a blast.

Use of the system is limited to our trained hook-up crews, since the complexity factor is high. A paper by my colleagues entitled "Improved blast control through the use of programmable delay detonators" will be published at the SEE in February 1997. This paper will present several case studies, and an analysis of the benefits of programmable detonators in certain applications, for the more technical reader.

This year, 1996, ICI Explosives acquired the global rights to this technology, and we are forging ahead with the commercial production of electronic blasting systems.

Experience with the first generation system has been incorporated into the design of the second generation system which is presently undergoing trials in South Africa. The new system will be released on a limited basis during 1997, and is a based on a simple 2 wire connection system. It is easy to use, substantially cheaper than the existing products and has been designed to address all mining applications.

Computer control of blasting is here to stay, and will soon be another building block in the modern computer controlled mining environment, networked and integrated with transport, drilling and ore processing control computers.

Blast Performance Management

James A. Hackett

ICI EXPLOSIVES EUROPE, ROBURITE CENTRE, SHEVINGTON, WIGAN, LANCASHIRE
WN6 8HT, UK

SUMMARY

This paper explores the potential to exploit controls over the outputs of the primary blasting operation arising from the performance consistency benefits gained in using modern bulk emulsion explosives technology. The former allied with the latest powerful blast predictive software and measuring tools offers the user the opportunity to apply control over both cost and performance which has not been possible in the past .

1. INTRODUCTION

There can be little argument that the pace of change in explosives technology since the 1800's to the present day has been quite massive and indeed continues at a significant pace even now.

Arguably since the early introduction of Blackpowder up to the Nobel era in the 1800's little development occurred. The pace thereafter has been brisk to say the least and especially since the 1950's major developments have occurred. Not the least of the developments in the latter era have been ANFO, Watergels and now Emulsions on the formulations side, while Shocktube and Transformer coupled detonators on the accessories side have all been major milestones.

Perhaps the area of most recent significant change in the UK has been the growth and development of computer software programs developed specifically for the explosives technologist and blasting engineer.

The combination of powerful performance predictive blast programs allied to the latest technology in bulk emulsion explosives opens up new horizons for the explosives user and indeed creates a fertile area for future development.

2.   BACKGROUND

In the past the presence of water in blast holes has necessitated the use of packaged high explosives over the more basic combination of high strength packaged base charge and ANFO where the holes were dry.   Apart from the obvious cost implications, variances in performance occur due to the thermodynamic effect of the residual water obliged to play its part in the reaction and also due to the fundamental characteristics of the different explosives which have to be used to meet these conditions.

Using a system of pumped explosives such as ICI's "Handibulk" emulsion explosives blend, which on loading into the blast hole fills its cross section and displaces any residual water, gives a level of performance consistency from blast to blast which has, largely, not been possible before on sites suffering from such variable conditions.   This consistency factor can be used as a vehicle in applying control to the process.

3.   DEVELOPMENT

If we consider the drilling and blasting operation as a process with key inputs and key outputs then there are in fact only three items on the input side of the overall equation. These are the explosives, the initiating system and the geometry which will be related to the specific site geology. Of these it is known that the explosives are the input item that has the least effect on the overall output performance, with the initiating system the second most important and the geometry that item with the greatest effect. This interrelationship is described in the blast performance triangle shown in **Fig1.**

When pumped bulk emulsion explosives are used then the operator has the ability to employ the same explosive with the same performance characteristics not only from hole to hole throughout the shot but time after time regardless of the residual ground conditions. This fact, allied with a cohesive quality control system, assuring the consistency of the product supplied to the blast holes and reinforced with dynamic testing such as continuous in-hole velocity of detonation checks, using ICI's "Powerline" VOD measurement equipment, provides the basis for control of the explosive.

As the initiating system itself is controllable both in its type and in its deployment then the focus of control and indeed the effect of any variance falls on the geometry of the blast layout. Obviously the geology of the mineral to be blasted is a significant factor which is specific to the particular site in question and therefore for the purpose of this paper is considered to be a constant input.

4.   INTERACTIVE MEASUREMENT

For the category of geometry the interrelationship between the blast holes is the key area of variance. Understandably, unless the holes are drilled in the same plane and at the same angle then variances in burden, spacing and even depth will occur.

The consequence of this is that the explosive may be asked to do more or less work than that planned in a particular blast hole according to its disposition to its nearest neighbours. This will result in variances in the fragmentation levels achieved across the blast with consequences in both dig rates and cycle times for the site operator .

Further downstream there will undoubtedly be effects on mineral throughput rates and power consumption in the materials processing area.

The effect of variances in burden and spacing can be modelled in terms of fragmentation distribution using interactive predictive blast models such as ICI's "Sabrex" blast performance predictive program.

By using face profiling techniques and a Pulsar Borehole Probe to measure the true extent of borehole deviation and misalignment for a particular layout the effects of maximum and minimum burdens and spacings against plan can be modelled.

The results of measurements taken on six blasts in a large quarry are shown in **Fig 2.** From these it can be seen that despite the nominal plan for this blast being a burden and spacing of 4.5m by 4.8m considerable variances were in fact measured when the compounded real values were measured. Minimum and maximum burdens were measured at 3.9m to 6.2m, while for the spacings 3.2m to 6.7m were measured.

These figures were used to predict three fragmentation distribution curves using the "Sabrex" program and these results are shown in **Fig 3, Fig 3a and Fig 3b.** As can be seen the effect is quite profound and although this prediction is a model, it is likely that in fact all these circumstances would be represented in different areas of these blasts and variances in dig rates etc. would be encountered throughout the mineral piles.

5. BENEFITS OF CONTROL

It should not be forgotten that significant variances in the alignment of blast holes which are drilled to a planned depth will result in holes which are short of grade with the consequence and further problem and cost of toes occurring and affecting the dig rate also.

The exertion and application of control will result in a consistency of fragmentation across the mineral pile and accelerate cycle times while at the same time reducing downstream handling costs. Wear rates and maintenance costs on prime movers will also be favourably affected.

It should be remembered that in most cases the drilling and blasting operation represents only 10% of the costs in a normal quarrying operation yet there can be little doubt that the quality of the execution of this function fundamentally affects the latter.

In order to exert control the limits of variance require to be determined and set. It should be borne in mind that in doing so the limitations of the drilling equipment in use need to be taken into account.

6.   SPECIFICATION DEVELOPMENT

The starting point for any program of development has to be a recognition of the status and performance variance of the current operation. In other words a benchmark of the competence or integrity of the standard operation is required.

This can be achieved by measuring the variances of the key elements of the geometry and ascertaining how these impinge on the performance of the blast.   Having gathered this data then it can be processed using the "Sabrex" model and manipulated until a representative picture of the standard shot in terms of fragmentation, heave and throw is developed.   For in depth or more advanced studies then fragmentation analysis techniques such as ICI's "Powersieve" software program, which relies on digital photographic studies of the mineral pile, can be used.   Other tools such as "Powerwave" which measures the velocity of the rock leaving the face on blasting might also be used.   Whatever, it is fundamentally important that as accurate a model of the standard working system as possible is constructed.

The consequence of the standard system on the downstream operation should be identified according to whatever bottleneck is believed to exist in the downstream process. This is likely to be in the load /haul cycle or the crushing area depending on the particular operation.

From this point the "Sabrex" model can be used to identify the limits of variance on the burden and spacing which are acceptable i.e. the Acceptance Criteria, in maintaining the required specification in terms of throw and fragmentation. These specification limits will require to be set for each blast hole in terms of depth, angle and azimuth.

7.   IMPLEMENTATION AND OPERATION

Measurement against the Acceptance Criteria which specify a blast hole is an active process whose intensity can only be defined according to the case in question.   Activity will certainly be at a high level during the early days of implementation; however, there is no question that this frequency can be lessened once control is assured. Thereafter a sampling regime may be applied providing that procedures and practices have been adjusted to meet the new regime.

Finally, it should be said that the acid test of any quality system such as that described is that if the controls are removed then the process should revert to its previous haphazard state and benefits should be seen to be lost.

8. CONCLUSIONS

The relatively recent introduction of bulk emulsion explosive blends into the UK extractive industry coupled with blast performance software and advanced measuring tools has opened up opportunities for the user to exert control over his blast performance. The benefits achieved are likely to be significant in terms of downstream process cost control and will be sustainable providing the Acceptance Criteria are correctly defined and maintained.

Note: Handibulk, Powerline, Sabrex and Powersieve are Trademarks of the ICI Group of Companies.

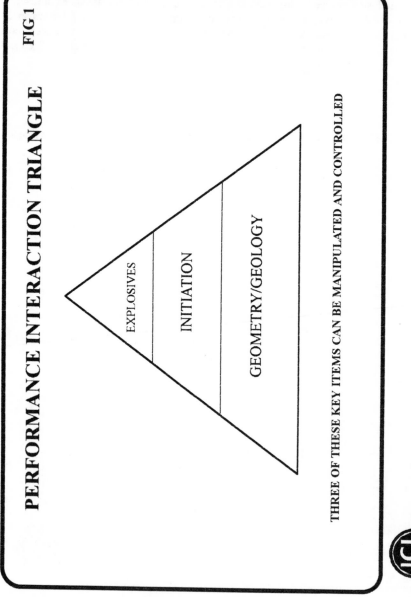

PERFORMANCE INTERACTION TRIANGLE   FIG 1

EXPLOSIVES

INITIATION

GEOMETRY/GEOLOGY

THREE OF THESE KEY ITEMS CAN BE MANIPULATED AND CONTROLLED

ICI Explosives

BENCHMARKING EXERCISE

FIG 2

127 MEASUREMENTS OF SPACINGS, 87 BORETRACKS AND 56 BURDENS BETWEEN 1ST AND 2ND ROWS.

BURDEN & SPACING RANGE

SPACING		RANGE	BURDEN		RANGE
MEAN	SPREAD		MEAN	SPREAD	
5.0	4.2 - 5.7	1.5	4.4	3.9 - 4.9	1.0
5.2	4.0 - 5.9	1.9	4.7	4.2 - 5.2	1.2
4.7	3.2 - 6.0	2.8	4.6	4.3 - 5.1	0.8
5.1	4.3 - 6.7	2.4	4.2	3.5 - 4.9	1.4
4.9	4.2 - 5.6	1.2	4.4	3.9 - 5.0	1.1
5.1	4.4 - 5.5	1.1	4.7	4.2 - 6.2	2.0

 Explosives

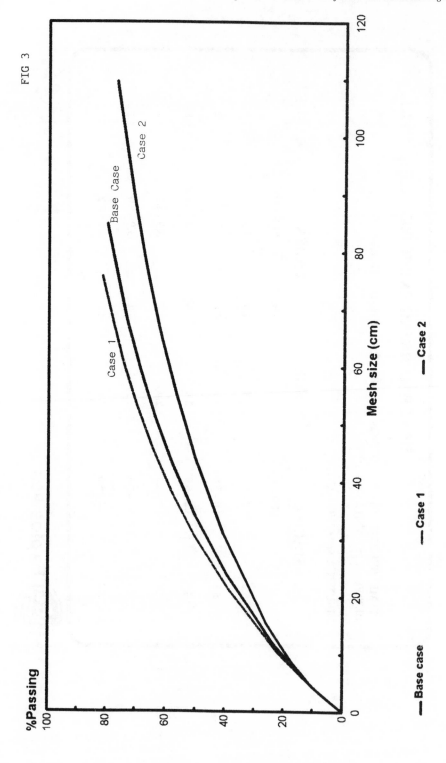

FIG 3

Design Summary

	Base case	Case 1	Case 2
Bench			
Height(m)	19.0	19.0	19.0
Face angle	75.0	75.0	75.0
Pit width(m)	22.0	22.0	22.0
Seam height(m)	0.0	0.0	0.0
Spoil angle	0.0	0.0	0.0
Grade dip	0.0	0.0	0.0
Pattern			
Type	Square	Square	Square
Drill dia.(mm)	127.0	127.0	127.0
Num. rows	1	1	1
Holes/row	0	0	0
Av. burden(m)	4.5	3.2	6.2
Av. spacing(m)	4.8	3.8	6.8
Av. PF(kg/m3)	0.636	1.130	0.326
Blast vol.(m3)	N/A	N/A	N/A
Blast mass(t)	N/A	N/A	N/A
Row timing			
Front row	?	?	?
Back row	?	?	?

Explosives in the Service of Man: The Nobel Heritage

FIG 3B

Results Summary

	Base case	Case 1	Case 2
Bobcat Factors			
Gr.lev.(GLF)	100	123	78
Col.frag(CNF)	100	119	82
Col.Blck(CBF)	100	125	74
Fly-rock(FRF)	100	110	85
Heave (HEF)	100	178	51
Heave			
Cr. X vel.	8.1	10.3	5.4
Cr. Y vel.	4.3	5.6	2.9
Mid X vel.	11.5	19.9	6.8
Mid Y vel.	5.3	8.4	3.2
Toe X vel.	11.1	19.0	6.6
Toe Y vel.	5.1	8.2	3.1
C.Mass disp.	16.3	32.0	7.6
Max. throw	30.9	54.0	18.7
% Cast	n/a	n/a	n/a
Fragmentation			
20% passing	10.1	7.0	15.1
50% passing	33.2	21.0	56.7
80% passing	81.0	47.8	150.7
Frag.index	7.9	6.9	9.1
Damage			
BackBreak	0.9	0.9	0.9

Controlled Blasting Operations and High Wall Stability in Surface Mining in Australia

Bruce C. Pilcher

ICI EXPLOSIVES EUROPE, ROBURITE CENTRE, SHEVINGTON, WIGAN, LANCASHIRE
WN6 8HT, UK (FORMERLY ICI AUSTRALIA)

1 INTRODUCTION

The profitability of most open cut mines and quarries is influenced by the slope of the final pit walls. Competition from low-cost producers in the Asia Pacific region has meant that an increasing number of Australian surface operations are dependent on steeper face angles and higher benches for their long-term viability. Steep stable pit walls can be formed by controlled or smoothwall blasting techniques which include cushion blasting, presplitting and postsplitting.

In some cases, stable pit walls can be created without smoothwall blasting. The blast design is the key to producing clean and safe pit walls at minimum cost. It needs to consider the rock conditions in the area, the likely amount of overbreak from a blast and the design location of the pit limit.

2 MODIFIED PRODUCTION BLASTS

Modified production blasts are a potentially low-cost method of forming stable pit walls. The modifications to a production blast include the optimisation of drill hole pattern, charge weight of explosives and the timing of the blasts immediately adjacent to the final wall to control the slope damage[1]. This method is mainly used in surface operations, e.g. the gold mines in the Kalgoorlie region, where the final walls are in weathered or oxidised rock which can be free dug to the pit limit.

Rock structure and strength, effective burden relief and the correct location of the blast holes are the most important factors.

2.1 Geology

Rock properties have a large effect on the ability of a production blast to create stable pit walls. Overbreak occurs where explosion gases can get into, wedge open and extend fissures in the rock and is generally greater where the rock fissures are closely spaced. Tight or in-filled fissures can cause less overbreak than open fissures. In closely fissured rock, the width of the overbreak zone is generally consistent and increases from the bench floor to bench top.

Medium to high strength rock with close jointing is an indication that stable walls may be produced from modified production blasts. Low strength, highly jointed rock types tend to overbreak excessively. Frequent overbreak greater than 1 or 2 burden distances indicates that even careful production blasts are unlikely to produce clean, stable pit walls[2].

2.2 Burden Relief

Effective relief of the burden during the blast has the greatest impact on the results of the final wall blast. Increased burden relief reduces the amount of backbreak produced by a production blast.

Blasting to a free face, minimising the number of rows in the blast and providing long delay times between rows of blastholes are three reliable methods of increasing relief and controlling backbreak. Reduction of the burden of blastholes is sometimes used to improve burden relief by more effectively throwing the rock away from the wall.

Angled blastholes along the crest of the bench may be needed to eliminate excessive toe burden as this is one of the major sources of poor relief during blasting. The back row of blastholes, nearest the pit limit, are generally drilled at an angle because vertical holes will penetrate the final wall and cause considerable wall damage (Figure 1).

Delay timing between rows is designed to allow sufficient time for the burden of each blasthole to detach itself completely from the rock mass before subsequent blastholes fire. When a back row blasthole fires, it will be able to heave its burden forward easily, causing little overbreak. Delay timing should also be chosen to reduce the number of blastholes firing at any one time and minimise the energy transmitted into the final wall.

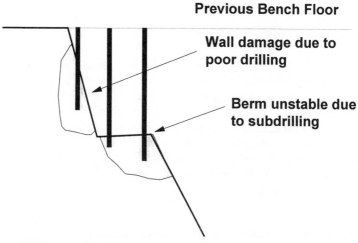

Previous Bench Floor

Wall damage due to poor drilling

Berm unstable due to subdrilling

Figure 1 *Wall Damage and Instability*

Stemming columns in the back row must be long enough to prevent cratering at the collar of the hole. Cratering from this row of blastholes will weaken the berm and lead to crest loss. Large craters are usually formed where the stemming length is less than approximately 60% of the effective burden distance. However, excessive stemming may cause the collar rock in front of the final limit to hang up, and result in a safety hazard.

2.3 Blasthole Location

The location of the back row of blastholes is critical to the location of the pit limit. The back row blastholes are drilled in front of the final pit limit to allow for backbreak behind the blastholes and the correct location depends mainly on previous experience in the pit, particularly if the amount of overbreak is variable.

If the stand-off distance between the back row and the pit limit is too small, there will be too much backbreak into the final wall. If the stand-off distance is too great, digging back to the design final face will be difficult and expensive. Backbreak is usually greater near the collar of blastholes than at the toe, so the location of the toe should be designed accordingly.

Subdrilling has to be carefully controlled when nearing final walls to ensure stability of berms below the current bench. Any subdrilling into a berm will reduce the rock strength and usually result in the loss of the crest of the berm, as shown in Figure 1.

3 CUSHION BLASTING

Cushion blasting is the simplest and least expensive smoothwall blasting technique; it is frequently overlooked when designing final wall blasts, but can be the most versatile and useful method of the three techniques. This method is successfully being used at Argyle Diamond Mine where presplit drilling has not been possible due to ground conditions[3].

3.1 Cushion Blast Design

The back row holes in a cushion blast contain lighter charges than the production blastholes and are drilled on a correspondingly smaller pattern. The diameter of holes are usually the same or sometimes smaller than the production blastholes in front of them.

Charge weight is commonly reduced by about 45% and both burden and spacing reduced by 25%. The energy factor is therefore essentially the same throughout the final wall blast. Cushion blastholes are delayed sequentially and detonated after the more heavily charged production blastholes in front.

Effective burden relief is one of the most important factors in the design of cushion blasts. As with modified production blasts, burden relief can be improved with: (i) a clean free face, (ii) a minimum number of rows, (iii) high powder factor and (iv) long delay times between rows assists in minimising back break into the final wall.

Cushion blasting is used alone where the rock is strong or where only minor reductions in overbreak are required. Minor overbreak and crest fracturing are invariably produced, even in the stronger rocks. Accordingly, cushion blasting is often used in conjunction with a more effective (but more expensive) smoothwall blasting technique such as presplitting. In this case, the cushion blastholes are located between the presplit line and the production blastholes.

Two rows of cushion blastholes can be used to help reduce overbreak. In this situation, the later firing row usually has a lighter charge than the earlier firing row. This technique is particularly useful in rock which backbreaks more than one burden distance.

Cushion blastholes with a smaller diameter than the production blastholes are usually more costly, but tend to produce sounder, smoother faces. Using smaller diameter holes reduces, but does not eliminate, the need for light charges corresponding to the smaller burden and spacing. The smaller burden and spacing of back row cushion blastholes increases the cost of drilling, priming, initiating explosives and blast crew labour.

3.2 Charging Cushion Blastholes

The energy distribution is carefully controlled to avoid overcharging by using stemming or air decks. Drill cuttings are most commonly used for stemming decks, but angular crushed rock is a more effective stemming material.

Air decks distribute the energy in cushion blastholes more effectively than stemming decks. The air space allows high pressure explosion gases to expand to a lower pressure before heaving the rock. Air decks can be formed in small diameter blastholes by blocking the hole with bags or a plastic spider which can be pushed to the appropriate level. The remainder is then stemmed. Large diameter blastholes can be blocked using an inflatable bag, Figure 2.

Decked charged blastholes reduce the charge weight so that the strain wave energy and volume of explosion gases are correspondingly reduced. However, overbreak is only decreased in areas of the rock face that also have reduced burden to break out.

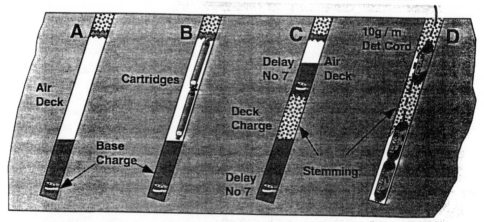

Figure 2 *Examples of Charging Cushion Blastholes*

3.3 Decoupled Charges in Cushion Blasting

Charges are decoupled when the charge diameter is smaller than the blasthole diameter. As the ratio of the blasthole diameter to charge diameter increases, peak blasthole pressure falls rapidly.

Decoupling is most suitable for reducing overbreak in closely fissured rock. Highly decoupled charges in the upper sections of back row blastholes can reduce crest damage considerably. In large diameter (i.e., 229 mm to 381 mm) blastholes, the effective density of ANFO can be reduced to as low as 0.06 g/cm^3 by decoupling the charge in PVC tubes. Figure 2 illustrates several methods of varying explosives energy distribution in cushion blasting.

4 POSTSPLITTING

A Postsplit or Trim blast consists of a row of parallel, closely spaced blastholes drilled along the pit limit. These blastholes are charged with a light, well distributed charge, and fired after the production blastholes have detonated. The postsplit holes split the rock between the blastholes to produce a sound smooth face with minimal overbreak. An indication of a good result is sighting the trace of the postsplit hole on the face.

4.1 Postsplit Design

To reduce costs, the diameter of the postsplit blastholes are the same or slightly smaller than the cushion and production blastholes. Larger diameter postsplit holes give better results because the effective deviation is less than smaller diameter holes.

Small and intermediate diameter blastholes can be charged with a string of small diameter cartridges of explosive suspended at intervals of approximately 1 metre on a detonating cord downline. However, to achieve optimum charge distribution, continuous columns of highly decoupled explosive should be used (e.g., 25 mm explosive for 89 mm diameter blastholes).

Burden and spacing of the postsplit blastholes should increase as the diameter increases, but the burden distance must always exceed the blasthole spacing (Table 1).

4.2 Combining Techniques

Cushion blasting is frequently used with postsplitting. The improved charge distribution in the cushion blastholes helps to minimise overbreak and increase the effectiveness of the postsplit.

In massive rock, postsplitting gives a considerable reduction in overbreak, but the final face is rarely as sound as that produced by presplitting. In closely fissured rock, on the other hand, the final face formed by postsplitting tends to be sounder than that produced by presplitting. The cost of postsplitting is lower than that of presplitting because the optimum spacing of the postsplit holes is larger than that for presplit holes.

Table 1 *Postsplit Design Guide for Average Strength Rock*

Diameter (mm)	Specific charge (kg/m)	Spacing (m)	Burden (m)
76	0.5	1.2	1.5
89	0.7	1.4	1.7
102	0.9	1.5	1.9
127	1.5	1.8	2.3
152	2.1	2.1	2.7
200	3.7	2.7	3.6
251	5.9	3.4	4.4
270	6.8	3.6	4.7
311	8.7	4	5.2

5 PRESPLITTING

Presplitting requires a row of closely spaced blastholes drilled along the design excavation limit, charged very lightly and detonated simultaneously before the blast holes in front of them.

Presplitting gives more spectacular results than postsplitting, but is generally more costly. This method is used in open cut mines, quarries and road cutting construction where the rock strength is moderate to very high with few joints. In most cases, the desired result after excavation is a clean face with the hole traces or 'half barrels' visible on the final wall.

The open cut coal mines in the Hunter Valley in New South Wales and the Bowen Basin in Central Queensland have recently introduced presplitting into the back row of overburden shots above the coal seam. Not only has presplitting assisted in maintaining highwall stability and maximising the volume of coal available for extraction, it has meant an opportunity to increase blast face heights from 35 to 60 metres.

5.1 Effects of Presplitting

Firing of the presplit charges splits the rock along the design perimeter of the excavation, producing an internal surface to which the later firing blastholes can break. The presplit plane acts as a pressure release vent for the explosion gases from the charges in the back of blastholes in front of the presplit. It also partially reflects the blast generated strain waves and so reduces vibration in the wall. The result is a relatively undisturbed face with minimum shattering, rock movement and overbreak. If the presplit blastholes are too close together or overcharged, they will produce overbreak[4].

Presplit blasthole diameters range from 76 - 102 mm in quarries and smaller open cut mine; whereas in the large pits, blasthole diameters up to 300 mm are now being used.

Presplitting may cause higher vibration levels than production blasts. The relatively high confinement of presplit charges may cause vibration levels per kilogram of explosive to be considerably higher than those for production blasts. Both the presplit and the subsequent adjacent blast should be designed to conform with the vibration limit established for the site.

5.2 Presplit Design

Presplit effectiveness depends greatly on good blasthole alignment. Hole deviation usually limits the length of 76 mm and 89 mm diameter blastholes to about 15 m. A major advantage of large diameter blastholes is that deviation is reduced and presplit blasts of greater depth can be fired. A single presplit blast may be fired for a number of benches. Where short benches are used (less than 6 m), presplits may be drilled 15-20 m deep to form a presplit for up to 3 benches. A single blast over 3 benches is cheaper and leaves a cleaner and safer final wall.

As blasthole diameter increases, the spacing between presplit blastholes normally increases, (Table 2). This table is only a guideline for designing presplits because rock properties have a dominant effect on blasthole spacing and charge load[5]. The optimum blasthole spacing and charge load for a particular rock should be determined by field trials. The depth of the hole will have a marked effect on presplit quality; holes that are

Table 2 *Presplit Design Guide for Average Strength Rock*

Diameter (mm)	Specific charge (kg/m)	Spacing (m)
76	0.5	0.9
89	0.7	1.2
102	0.8	1.3
114	1.1	1.4
127	1.3	1.5
152	1.9	2
200	3.3	2.6
251	5.3	3.3
270	6.1	3.6
311	7.8	4

too deep may cause overbreak, but short holes will give insufficient cracking. This may require extra toe holes for excavation and the careful splitting effect would be lost.

5.3 Charging Presplit Blastholes

In average rock conditions, the charge load required for effective presplitting increases with the blasthole diameter, as shown in Table 2. Optimum charge load varies considerably with rock properties[6]. Very weak or closely fissured rock needs a reduced charge load and blasthole spacing. Massive rock with high dynamic tensile breaking strain could require a higher charge load.

In unconsolidated ground, the charge weight per linear metre in the upper portion of the blasthole may need to be reduced by 50% or more to minimise overbreak at the crest.

In Australia, quarries and open cut gold mines use hole diameters 76 - 102 mm for presplitting. The presplit blasts are loaded and fired with Powershear® and 10 gm detonating cord. Powershear® is a continuous small diameter cartridge explosive which is available on reels to speed up the charging process. Some excellent results have been experienced in quarries which require presplitting of vertical walls around a primary crusher[7].

Continuous charges ensure energy is distributed evenly along the blasthole. Energy concentration of continuous charges can be varied by taping two or more continuous charges together, or by changing the stemming length or unchanged collar length of the blasthole. Changing the blasthole diameter will vary the effective energy concentration.

Presplit blastholes should generally be charged to within approximately 8 blasthole diameters (d) of the collar. In closely fissured rock, the uncharged collar may need to be as much as 15d.

In large diameter holes, say 200 mm and over, there is no specialised explosive available. The most common method is the combination of airdecks with a bulk emulsion or ANFO in the base and the column of the hole[8]. Large diameter packaged explosives placed at regular intervals in the hole and connected to a detonating cord downline is sometimes used.

5.3.1 Water-filled Presplit Holes. Water is incompressible, so the explosives energy from decoupled charges is more effectively transmitted and driven into the surrounding rock. Existing joints will be wedged open by the powerful hydraulic force of the water. In solid, massive rocks, not many new cracks will be formed, so a good presplit can be expected. Highly jointed rock will be damaged and loosened more than an equivalent dry hole presplit.

5.3.2 Stemming Presplit Blastholes. Presplit blastholes should not be stemmed unless there is a need to control airblast. Leaving the collar of presplit unstemmed allows the explosion gases to jet into the atmosphere very rapidly. This ensures they are less likely to jet into cracks which intersect the wall of the blasthole. Unstemmed holes reduce damage to the crests of final berms and lower the probability of crest loss.

The spacing of stemmed presplit blastholes can be increased, because the confined explosion gases help the propagation of the presplit. Unfortunately, stemming also increases damage to the crests of final berms, Figure 3.

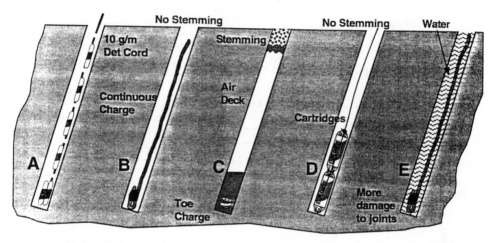

Figure 3 *Examples of Charging Presplit Blastholes*

5.4 Firing Presplit Blast

Presplit charges must be initiated simultaneously by joining all downlines from the presplit blastholes to a detonating cord trunkline. Where ground vibrations are likely to cause overbreak or disturb residents, delays are used to fire groups of blastholes. The number of blastholes in each group should be sufficient to achieve a satisfactory splitting action while not exceeding the maximum charge weight that can be fired per delay.

If detonating cord trunklines cannot be used, each downline can be initiated at the collar of the blasthole by a short delay detonator (preferably the zero delay).

Presplit blastholes should be fired in the same blast as the adjacent final wall blast if the total burden distance is smaller than about 150 times the diameter of the presplit blastholes. If the burden on the presplit blast is less than this, ground movement could cause cracking of the rock. The broken ground can cause drilling, charging or cutoff problems.

Best presplit results are generally obtained when presplit blasts are fired separately and ahead of the adjacent final slope blast. Separate firing is possible if the total burden distance is very large or when blasting in solid rock. The burden on the presplit blastholes needs to be sufficient to prevent movement of the entire block of rock in front of the presplit.

5.5 Protecting the Presplit Face

Presplit faces will be damaged if production blastholes are drilled too close. On the other hand, if the stand off distance between the presplit and the blastholes in front of them is excessive, unbroken rock will be left on the presplit face.

When correctly drilled, the back row charges of the final wall blast fragment the rock in front of the presplit. Optimum stand off distance can be determined only by trials and is usually 40 - 70% of the burden distance of the back production (or cushion) blastholes.

5.6 Current Research

ICI Explosives is continuing to conduct research on suitable design tools and explosives systems to optimise final wall blasting results, particularly with large diameter blastholes and to assess the quality of presplits[6].

Acknowledgements

I wish to thank ICI Explosives for permission to publish this paper. My colleagues in Australia and the U.K. are also thanked for their advice.

References

1. E. Hoek & J.W. Bray, 'Rock Slope Engineering', Revised Third Edition, Institute of Mining & Metallurgy, 1981.
2. A. Scott, 'Damage to Open Pit Slopes', Third Large Open Pit Mining Conference, Mackay, 30 August - 3 September 1992.
3. B.M. Bulow & J. Chapman, 'Limit Blast Optimisation at Argyle Diamond Mines', Open Pit Blasting Workshop, Perth, September 1994.
4. J. Gamble, 'Overbreak Control with Large Diameter Blastholes', The Australian Coal Journal, No. 39, 1993.
5. ICI Australia, 'Safe and Efficient Blasting in Open Cut Mines', ICI Explosives Technical Services, May 1996.
6. G. Brent, 'The Design of Pre-split Blasts', Explo '95 Conference, Brisbane, September 1995.
7. A. Brodbeck, 'Final Wall Blasting - A Case Study', Quarry Australia Journal, Institute of Quarrying, March 1995.
8. P. Dunn & A. Cocker, 'Wall Control for Surface Coal Mines', Explo '95 Conference, Brisbane, September 1995.

Session 7. Blasting Practice

Marine Blasting for Rock Excavation

P. C. Francis

ROCK FALL COMPANY LIMITED, WESTMINSTER HOUSE, CROMPTON WAY, FAREHAM, HAMPSHIRE PO15 5SS, UK

1 INTRODUCTION

In the early part of this millennium man made use of natural harbours, bays and estuaries to provide safe anchorage for his vessels.

As time went by the size of ships increased as did man's ability to construct harbours, breakwaters and quays. This led to the development of dredgers that could remove silt and sand to give sufficient water depths. With the increasing size of ships it became unusual for the dredging to be carried out without encountering rock outcrops or obstructions that would have to be removed.

It is with thanks to Alfred Nobel that some of the earlier channels could be cleared with the use of explosives. It was, as we know, in 1862 that Alfred Nobel successfully exploded nitroglycerine underwater in his factory canal in Stockholm. This was a year before Nobel found a safe method of detonating nitroglycerine by means of a detonator which he patented in 1863.

The availability of suitable dynamite, and later development of Blasting Gelatine for underwater use, made previously impossible underwater operations feasible. Examples are:

- the rocks at Hellsgate in New York's East River 1876;
- the Corinth Canal (1881-93);
- the clearing of the Danube at the Iron Gates (1890-96).

All these projects were successfully completed with the assistance of explosives.

In the construction of the Panama Canal (1880-1914) the majority of the canal was drilled and blasted in dry conditions. However at the breakthrough to the sea at each end of the canal the work had to be carried out by marine methods.

There are few reports of activity, other than the Panama Canal, in commercial underwater blasting in the first half of the 20th century. However, in the early 1950s Atlas Copco and Skanska, both of Sweden, introduced the forerunner of the overburden drilling machine, known at the time as the LINDO RIG, which was built for the particular purpose of drilling and blasting rock through heavy topsoils to form the Lindo Canal in Sweden.

The system is based on a rotary percussive drill which can drill casings through several metres of earth into rock, then drill a hole via the casings into the rock and finally after the hole has been charged with explosive, retrieve the casing. It was quickly appreciated that if the earth was replaced by water, then marine drilling and blasting could be carried out from a floating or fixed platform without the use of divers working on the seabed.

Ports and waterways continued to be constructed, enlarged and deepened. This was due to the development of third world countries, the growing demand for oil as the major source of motive power and the general increase in world trade.

To keep pace with these requirements larger, more sophisticated and powerful dredgers were developed enabling greater dredging depths to be achieved and with increased output.

Invariably the deeper the dredging depth became, the more rock was encountered. Although some of the larger dredgers are now able to remove soft and medium rocks directly, the harder rocks still require some method of pre-treatment.

This presentation is intended to cover the practical use of explosives to fragment rock for dredging operations connected with harbour deepening, navigational channel clearance, quay wall construction and foreshore trenching. Most of this work is associated with the Dredging Industry.

2 DREDGING

The Dutch, due to the need to safeguard and develop their low lying country, became the pioneers of the dredging industry. Dredging started with men working from small boats which they loaded with long handled spades or scoops. Man's ingenuity, however, eased this arduous task through the development of mechanised digging systems.

The first system consisted of scoops fixed on the end of wooden spokes which were rotated by hand from the deck of a small boat,scraping the soft material from the river bed and depositing it into an adjacent craft. This method was obviously limited to shallow waters and the depth of dredging could not be altered to allow for tides and local river bed conditions.

The system then developed to a series of buckets driven by a man-powered treadmill. This was the forerunner of what, for many years, was the main work horse of the dredging industry, the bucket dredger.

Figure 1 *Bucket Dredger*

2.1 Bucket Dredger

The bucket dredger was at one time the most universally used dredging tool, capable of dredging sand, silt, stiff clays and soft rock. This type of dredger is a floating pontoon, fitted with a long arm known as the ladder which can be lowered to the sea bed. The ladder, has a continuous chain which passes around it. Attached to the chain is a series of specially designed buckets which have a cutting edge or teeth along the front lip. When the ladder is lowered to the seabed and the chain set in motion, each bucket cuts and scoops up the material which is tipped onto a chute into a barge moored alongside.

Although many bucket dredgers can still be found working around the world they have in general, been superseded by other types of dredgers and there are now three main categories of dredger suited for dredging blasted rock.

2.2 Cutter Suction Dredger

This dredger is used mainly for dredging sand, clay and soft rock materials, not only to deepen harbours and waterways, but also to reclaim land from the sea.

As its name suggests, the operation is one of cutting the seabed materials to loosen them and then sucking up the materials and hydraulically transporting them to the deposit area.

In principle the dredger consists of a floating platform, at one end of which is a pivoted arm that is lowered to the seabed. At the end of this arm is the cutter head which has rows of teeth (pick points) and the head is rotated in order to cut and loosen the seabed materials. Behind the cutter head is the suction mouth connected to a pipe and the dredged material is passed through the pump via a pipeline to the deposit area.

These vessels vary in size from having installed pump capacities ranging from 1,000 hp to 12,000 hp and pipe diameters between 30 cm and 100 cm. The most powerful dredgers can cut rock of hardness up to 20 - 30 MPa without any pre-treatment.

In general the majority of cutter suction dredgers are stationary vessels. However, recently, a new generation of large, sea-going, self-propelled cutter suction dredgers have been built.

2.3 Backhoe or Backacting Dredger

The large backhoe dredger is ideally suited for dredging underwater blasted rock and particularly for sub-sea trench dredging. However, it has its limitations in dredging depth.

The dredging tool is a large hydraulic backacting excavator, much the same as is used in the construction industry, and is mounted on a turret instead of tracks. This unit is installed on a purpose-built spudded pontoon. On occasions, tracked land excavators have been installed on a flat top barge to carry out dredging works.

As with land excavators, hydraulic power has superseded the old rope operated machines, but the excavating operation remains the same.

Material dredged by this method is usually placed in a hopper barge alongside the dredger or cast to one side for re-use.

2.4 Grab Dredger

The grab dredger can operate in deeper water depths and can dredge most materials including blasted rock.

It is rope operated in the same way as a land based grab excavator, as can be seen unloading bulk grain from ships for example.

Again, as for the backhoe dredger a crane similar to those used on land works is mounted on a turret on a floating pontoon. The crane is fitted with a rope operated grab and has a large heavy bucket with teeth to tackle the more cohesive soils and rock. For rock excavation the weight of the bucket is very important, the weight is required to achieve penetration into the material to be dredged. If the grab is too light it will not dig into the material and the bucket will not penetrate the rock pile.

Materials dredged in this way are placed in hopper barges for depositing elsewhere or cast to one side for re-use.

3 MARINE ROCK BLASTING

Although there are non-explosive methods of fragmenting rock underwater to assist dredging operations, such as: hydraulic breakers, chisels and expanding chemical compounds, 99% of rock fragmentation is carried out with the use of explosives.

There are two blasting techniques available, these being surface blasting and drilling and blasting. Surface blasting as its name suggests is carried out by laying explosive charges in direct contact with the rock surface, whereas drilling and blasting requires holes to be drilled in the rock which are column-charged with explosives.

4 SURFACE BLASTING

The method of laying explosives charges, either in a boxed form or as several sticks bound together and placed in direct contact with the rock, requires a very high explosives : rock ratio (ie the blast ratio) compared to that which is required in drilled shotholes. This is beneficial for the explosive manufacturers but not too good for the environment.

This method is usually considered when drilling and blasting is not economical or practical, for example when:

- water depths are too great for the drilling rig;
- the rock is in thin layers overlying soft material, with no overburden present;
- the cost of mobilising a drill barge for the quantity of rock to be removed is prohibitive.

The main advantages of surface blasting are that no drilling plant is required and a large number of charges can be quickly prepared and placed in position in a short period.

Surface blasting depends entirely on the shattering effect of the detonation wave and the gases produced are dissipated in the water.

The blasting of solid rock by surface charges has only a limited effect. It is of little use in basalts and granites, but is useful for fragmenting thin layers of sedimentary rocks which over-lie softer materials.

In an attempt to improve the efficiency of surface blasting the "shaped charge" was developed. This is a conical canister which can be filled at the factory with water gels or emulsion type explosives.

The main purpose of the shaped charge is to direct the detonation shock downward into the material to be broken giving a greater efficiency of blasting.

It is very important that the charge is placed in direct contact with the rock. If this is not achieved there will be little advantage over using "boxed" explosives.

This method can be considered only in areas where there is little or no overburden overlying the rock or where the overburden can be cleared prior to blasting. Several metres of water are preferable for surface blasting as the efficiency of the explosives charges is increased by the tamping effect caused by the pressure of the water.

Where large areas have to be covered a frame can be utilised. When using a frame charges are secured to it in a set pattern, usually on the deck of a crane barge, after which the frame is lowered to the sea bed. The charges are then released from the frame by divers or mechanically by a remote device, the frame is raised and the barge is winched to the blasting station where the safety checks are completed prior to blasting.

The effects on the environment restrict surface blasting operations. The unconfined charges release energy in all directions, with large amounts of the energy being lost immediately to the water. The resulting waterborne shockwave is very great. In shallow water there is a large waterspout, accompanied by an airborne shockwave. Heavy ground vibration also results and therefore the method cannot be used where ground vibration control is required.

5 DRILLING AND BLASTING

Underwater drilling and blasting techniques can be compared to land blasting operations, except that with underwater blasting the results cannot be seen. There is no way of knowing that the blast just fired has successfully broken all the rock to the required depth. This will only be evident when the dredging operation is complete.

There are some very important differences between land and marine operations the result of which is a substantial increase in the explosives blast ratio, which is measured in kilograms of explosive per cubic metre of rock blasted.

The technique requires holes to be drilled to predetermined regular patterns, charged and detonated to provide fragmentation for different types of dredging equipment. Bench heights are relatively low, holes are vertical and the face will not normally have been cleared prior to the blast. Therefore, the techniques are more closely related to crater blasting than conventional free-face blasting.

The blast ratios are very much higher than for land operations because:

- the operation is basically a cratering one;
- the pressure of the water and overburden contains the blast and restricts the heave and throw of the rock;
- previous blasts will not normally have been dredged, thus further confining the rock movement;
- secondary blasting must be avoided due to the degree of difficulty in carrying out this work and therefore the high cost of re-drilling;
- normally holes have to be drilled vertically, denying the possibility of using the benefit from angled holes which improves the throw.

Over-drilling or subgrade drilling below the designed dredge level is necessary in underwater blasting to ensure fragmentation is achieved to design level.

Initiation of underwater blasts can be by long-lead short delay electric detonators, non-electric shockwave tube system and by detonating cord. Detonating cord is less frequently used nowadays due to the adverse environmental effects. The improvement in strength of non-electric shock tubes with millisecond delay detonators and their safe application in the blasting process has led to an increase in use of this system.

The nature of underwater drilling and blasting work varies significantly from contract to contract and can be carried out in water depths of up to 35 metres.

Several factors influence the execution of the work, in particular:

- the nature and formation of the rock to be blasted;
- overburden types and the thickness;
- the degree of fragmentation required for the type and size of dredger to be used to excavate the rock;
- vibration restrictions on nearby structures;
- environmental effects on water sports, swimmers and divers;
- the depth of water and tidal range;
- range of swell and currents.

All the above are factors which should be considered at the estimating and planning stage.

When working in operational harbours the sea conditions are usually more favourable, but other delays, such as shipping movements, require to be taken into account in the planning, and it is more often than not the contractor's responsibility to take full account of these delays and allow for them in his rates and programme.

Contracts can vary in size from the drilling and blasting of a few cubic metres (single boulders/small outcrops) to large scale bulk drilling and blasting works in the order of hundreds of thousands of cubic metres.

Although drilling and blasting can be carried out by divers with seabed equipment it is normally carried out from the sea surface, with the drilling equipment mounted on floating barges or jack-up platforms. The floating barge would usually be equipped with an anchor winch system to hold the barge steady whilst drilling and to provide good manoeuvrability for positioning.

The size of the drill barge and the number of drill towers is dependent upon the size of the job and the time required to complete the operation, the optimum number of rigs being established at the estimating stage.

Drill towers can be independently moveable along the edge of the barge or work from a well in the centre of the barge, enabling the rig to drill a number of holes without re-positioning the barge.

Figure 2 *Drill Barge*

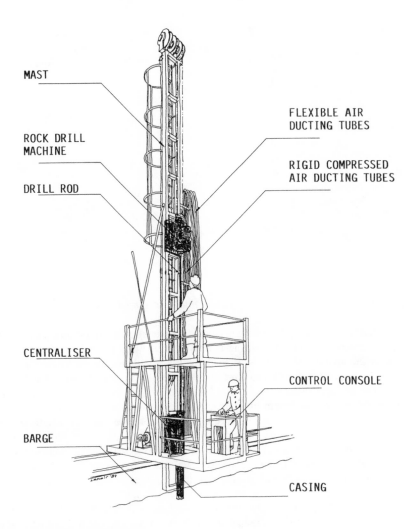

MAST

FLEXIBLE AIR
DUCTING TUBES

ROCK DRILL
MACHINE

RIGID COMPRESSED
AIR DUCTING TUBES

DRILL ROD

CENTRALISER

CONTROL CONSOLE

BARGE

CASING

Figure 3 *Sketch of a Drill Rig*

There are various methods of drilling such as "Stand Pipe", "Kelly-bar" and "OD". The OD method of drilling is widely used for marine operations.

Figure 3 shows a sketch of a typical drilling rig working over the side of a flat top barge.

The drill has a rotary percussive action and can drive the casings and the drill rods as necessary. It can be powered by either compressed air or hydraulics. The height of the mast is usually some 15 - 30 metres and the casings vary between 100 mm and 150 mm diameter.

The drilling process is started by building up the outer casing tubes until they reach the seabed. The casings are then driven through the overburden until they are collared into the sound rock forming a seal for the top of the hole.

The inner drill steels are then built up and the hole drilled supported by the casings.

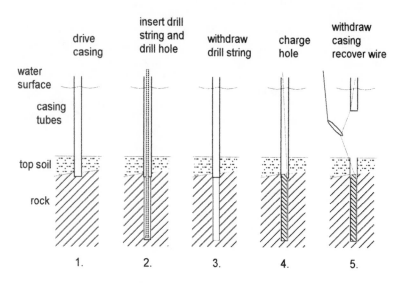

Figure 4 *Schematic Representation of Drilling and Charging Cycle*

On completion of drilling, the hole is checked for irregularities and thoroughly flushed to remove any chippings. The drill string is then removed and the hole is charged.

As part of the drilling cycle the rock head level is calculated for each hole by noting the length of casing between rock head and the water level, taking into account the level of the tide. The difference between this level and the dredge design level give the depth of hole to be drilled and the length of charge required.

Once the charge length is known, a plastic tube is cut to the required length and one end is closed in the form of a cone, the other being left open. A primer charge is made up by inserting a long-lead millisecond delay detonator into a cartridge of explosives, care being taken to use the correct delay number selected according to the blast pattern.

Further cartridges of explosives are inserted into the plastic tube to make up the full charge, and the open end then sealed with tape.

Once the hole has been completed the prepared charge is lowered into the hole and the level of the charge checked by graduated stemming rods. The casings are then raised and the detonator wire recovered using a rope ring which is placed round the casing and then lowered by rope until it passes the end of the casing when it is pulled to the surface with the detonator wire. The end of the wire is then secured to a convenient point on the barge until required for connecting up prior to blasting.

The rig is then moved along the barge to the next drilling position and the cycle repeated until a row of holes is complete. Following completion of a row of holes the barge is moved to the next row position and the drill and charge cycle continued until the blast plan is completed.

On completion of the blast plan the individual detonator wires are connected to brass buss bars, which are fixed on a substantial float, to form a parallel circuit. A low resistance cable is then attached and the float is lowered into the water and the barge winched to the predetermined firing station. The cable is attached to an exploder and blasting will then take place after the warning procedures have been completed.

5.1 Blast Ratio, Fragmentation, Bulking

The blast ratio is the quantity of explosives used to blast each cubic metre of rock and is calculated by dividing the total charge placed in the hole, by the theoretical volume to be blasted, including the sub-grade volume.

Sub-grade drilling is the depth drilled below the design dredging level to ensure that blasted rock can be removed to the required level. This is necessary as the dredger must penetrate into the sub-grade in order to leave the rock at or below the design level.

The amount of sub-grade drilling is usually equal to the distance between bore holes so if the drilling pattern is say, 2 metres x 2 metres, then 2 metres of extra drilling and charging will be required.

This extra drilling is what we call "unpaid work" and it will dramatically increase the amount of explosives required and therefore the cost of blasting, particularly when the thickness of rock to be removed is less than 1 metre.

Table 1 refers to the "Total Blast Ratio" and is intended to apply for custom-built rock dredgers when using 80% weight strength nitroglycerine explosives which are initiated by millisecond delay techniques.

A combination of rock type and dredging equipment must be considered in order to select the drill pattern which will give a blast ratio that will result in the rock being sufficiently bulked and fragmented to allow efficient dredging whilst controlling costs by not over-blasting.

Total blast ratio can be defined as the total explosives required to blast the full depth of hole including sub-grade drilling. The total volume of rock blasted should also be used in the calculation, not just the rock above design level.

The results of the drilling and blasting will depend mainly on the rock type and its characteristics. This will determine the size distribution, and the bulking resulting from the blast. A combination of the hardness, friability, brittleness, degree of weathering and whether the rock is igneous, sedimentary or metamorphic will therefore determine the blast ratio required to achieve the desired fragmentation and bulking for the specific dredger.

Table 1 *Typical Blast Ratios*

Description	Type of Rock	Type of Dredger			
		Bucket	**Cutter**	**Backhoe**	**Grab**
		kg/m3	kg/m3	kg/m3	kg/m3
Easy to blast	**Corals, medium limestones, mudstones, shales, siltstones**	0.60	0.60	0.55	0.80
Moderate to blast	**Hard limestones, sandstones, marls**	0.80	0.90	0.75	1.00
Difficult to blast	**Granites, dolerites, gneisses, basalts**	1.00	1.20	0.90	1.20

Table 2 *Distribution of the Size of Rock after Blasting*

Dredger Type	Relative Digging Force	Degree of Bulking Required	Maximum Size Distribution Required
Sea-going Cutter Suction Dredger	Very High	20%	100% < 500 mm 90% < 300 mm
Backhoe Dredger	High	20%	90% < 1000 mm 70% < 500 mm
Bucket Dredger	Medium	30%	90% < 500 mm 70% < 350 mm
Grab Dredger	Low	40%	90% < 1000 mm 70% < 350 mm

Table 2 gives an indication of bulking and size distribution required for the four main types of dredger.

The degree of bulking required is the increase in volume of the rock pile resulting from voids in the rock after blasting. The fragmentation required depends mainly on the breakout force which the dredger can apply to the blasted rock.

A large cutter suction dredger for example can apply very large digging forces and remove rock which has been pre-treated with a blast ratio less than that which would be required for a grab dredger. The grab dredger requires the rock to be well bulked in order that the bucket can penetrate, but is able to remove much larger rock fragments than those which would pass through the pump of a cutter suction dredger.

6 VIBRATION CONTROL

Groundborne vibration is the dominating factor when assessing the risk of underwater rock blasting. When the work is within harbour areas it is very often adjacent to waterlogged and infilled areas and the blast vibrations are more easily transmitted through this type of material.

Particle velocity is the most commonly used measurement to determine the threshold of damage. Peak Particle Velocity (PPV) is expressed in mm/s.

Generally the maximum levels of vibration applicable to nearby properties and structures are set out in the contract specifications. However, in their absence, maximum PPV limits have been established in relation to various types of structures and properties. Table 3 shows a limit of 15 mm/s at 4 Hz increasing to 20 mm/s at 15 Hz for residential properties. It is wise to keep this level below 5 mm/s on occupied properties so as to reduce the nuisance factor.

After initial calculation of predicted PPVs trial blasting is first carried out at the blast area, using smaller charges than those planned for the main operation. The vibration information gathered enables the calculation of the site parameters, which establish   the relationship between charge weight distance and resultant PPV. Maximum charge per delay can then be determined for different distances from the source of concern to maintain vibration levels within safe limits.

Table 3 *Transient Vibration Guide Values for Cosmetic Change*[1]

Type of building	Peak component particle velocity in frequency range of predominant pulse	
	4 Hz to 15 Hz	**15 Hz and above**
Reinforced or framed structures Industrial and heavy commercial buildings	50 mm/s at 4 Hz and above	
Unreinforced or light framed structures Residential or light commercial buildings	15 mm/s at 4 Hz increasing to 20 mm/s at 15 Hz	20 mm/s at 15 Hz increasing to 50 mm/s at 40 Hz and above
Note: Values referred to are at the base of the building		

7 WATERBORNE SHOCKWAVES

Consideration must also be given to waterborne shockwaves, and airblast, particularly at close proximity to quay walls.

The important parameters of the shockwave are the peak overpressure and the impulses. Much work has been done in the measurement of overpressures resulting from freely suspended charges but less so for explosives placed in a drilled shothole. However, various studies indicate the shockwave pressure from a shot buried in a drill hole is only 5-10% of that expected from a freely suspended charge of the same weight.

The effect of the shockwave overpressure on living creatures varies considerably. It has been shown that crabs, lobsters and oysters are relatively unaffected and are able to withstand fairly high shock pressures. Fish with well developed air bladders are more affected than those with smaller bladders.

Swimmers and divers too close to an underwater blast can suffer injury and possible death. Peak overpressure is the parameter generally related to injury but some sources indicate that the impulse is also important.

Table 4 is based on freely suspended TNT explosives. Considerably less water shock is produced when charges are contained in boreholes.

Table 4 *Safe stand off for divers and swimmers*[2]

Size of charge	Minimum distance
kg	**m**
Up to and including 10	600
Over 10 and up to and including 20	750
Over 20 and up to and including 30	900
Over 30 and up to and including 40	1050
Over 40 and up to and including 50	1200

The overlying water present in underwater blasting operations substantially reduces airblast. Even so, airblast is still probably the biggest cause of complaints arising from such operations. Particular attention must be given to it when using surface blasting techniques. Surface blasting is impracticable within, or close to, quiet areas such as residential suburbs, hospitals, etc.

The effect of waterborne shockwave can be reduced by the use of an air bubble curtain; however, except in very limited applications, the quantity of air required and the difficulty in laying the necessary pipes, hoses etc limit the use of air bubble jet curtains.

8 CONTROLLED BLASTING TECHNIQUES

A very practical and useful method of reducing vibrations and waterborne shockwaves is to reduce the maximum charge per delay by the use of deck charging techniques. This is done by placing various charges, each with a different millisecond delay detonator in one hole. The individual charges are separated within the plastic tube used when making up the charges by layers of stemming (usually in the form of gravel) to avoid sympathetic detonation.

Care has to be taken when using this technique with the selection of the type of explosives, the hole diameter and the distance between individual charges as these factors can all cause sympathetic detonation when using gelatinous explosives, and desensitisation when using emulsions. When this work method is to be used it is advisable to carry out initial trials at sufficient distance from structures such that should sympathetic detonation occur the vibration limit for that particular structure is not exceeded. This technique is generally only used when working within 20 m of existing structures.

In cases where it is necessary to control over-break to form vertical faces or protect structures, pre-splitting or line drilling techniques can be used. Pre-splitting is carried out prior to bulk blasting and involves the drilling of a series of holes at close centres along the required design line. The holes are then charged with small diameter explosive. The maximum number of holes which can be fired without exceeding the vibration limit are then blasted simultaneously. This results in a vertical fracture in the rock between the holes which will limit over-break when subsequent blasting of the bulk rock is carried out.

Line drilling is used to created the same effect as pre-splitting but is utilised in areas where the vibration limit would be exceeded by blasting the pre-split holes. Line drilling is carried out by drilling holes in a line at very close centres. When subsequent bulk blasting takes place the rock will fracture along the line of these holes preventing over-break.

The results of applying these techniques to dry land operations can be seen in many of the rock cuttings which have been formed in recent years for the passage of motorways.

9 CONCLUSION

From the preceding information it can be seen that world trade has been greatly influenced and enhanced by the successful development of underwater explosives initiated by Alfred Nobel.

References
1. BS 7385:Part 2:1993 *Guide to damage levels from groundborne vibration*
2. BS 5607:1988   *Safe use of explosives in the Construction Industry*

Tunnel Blasting – Recent Developments

T. E. White

EXCHEM EXPLOSIVES PLC, COMMONWEALTH HOUSE, NEW OXFORD STREET, LONDON, UK

INTRODUCTION

In spite of the development and growing use of tunnel excavating machines of all types and sizes, explosives continue to be the tool which enables the majority of the world's tunnels to be driven efficiently. The history of the use of explosives is now three centuries old and has undergone many changes. The rate of change has increased in the past quarter of a century or so. In the last century black powder has given way to the use of dynamites and gelignites. Dynamites and gelignites are yielding to the superior benefits of the modern explosives, slurries and emulsions, both based mainly on the use of ammonium nitrate solutions. Ammonium nitrate and fuel oil itself has been considered for almost every blasting application, including tunnelling, though it has had to be dismissed on many occasions because of its low bulk strength and lack of water resistance.

Fuses were developed to fire black powder, and fuses assembled with plain detonators were used with gelignites and dynamites until they were replaced by delayed action detonators - either long or short period delay.

Shock tube systems used with modern explosives give a degree of safety which has enabled improvements in performance and production to be obtained. The increasingly sophisticated drilling machines have done much to reduce the cycle time and to improve the rate of advance of modern tunnels of all sizes.

Further refinements, such as the introduction of the much awaited electronic detonators and a greater use of bulk explosives, should guarantee that where conditions are suitable, drilling and blasting can remain a viable option in major tunnel drivages for a long time into the future.

TUNNELLING MACHINES VERSUS DRILLING AND BLASTING

Full face tunnelling machines achieve some remarkably high rates of advance where conditions are suitable. Improvements in the design of these sophisticated pieces of equipment have made them more versatile. Not many years ago, one was only confident that such machines would succeed in soft weak rocks which contained no abrasive material. Of course it was always helpful if the tunnel was long and straight. The preparation underground of a length of tunnel in which the machines were to be built usually meant that the project commenced with an excavation assisted by drilling and blasting.

A similar station is necessary at the completion of the drive so that the machine can be dismantled (if the machine is going to be salvaged at all - some have been buried and abandoned after completing their task). The need to support incompetent ground when it is encountered, and the problems produced by other unanticipated geological anomalies are always a concern. Full face machines can now cope with much tougher conditions and strong hard rock containing abrasive material presents much less of a problem.

However, tunnelling machines of all types are very expensive and are inflexible in application. Initial high rates of advance soon tumble from hundreds of metres a week to a few centimetres a day when the machines are challenged by adverse conditions. Once a machine has started on its way there is not a great deal one can do except wait for it to arrive at the terminus. There are not many options for change.

If the tunnel is long, level, straight and circular, passing through competent though not tough and abrasive rock, full face tunnelling machines and road headers are the obvious choice.

If the tunnel is short, small, other than circular in shape or the rock is changeable or unproved, excavation by drill and blast is a serious consideration. If additional access points can be effected, other than at each end of the tunnel, mobile drilling machines, excavators and suitable haulage systems usually guarantee that even when difficult ground conditions are encountered, progress is being made at least on some faces with some time recovery.

Tunnelling machines of the cutting head type are a little more flexible than full face machines and in some situations it is possible to combine drill and blast with machine excavation. In such cases, the tunnelling machine drives through the softer measures until either lack of progress, excessive wear due to abrasion or a combination of both precipitate the withdrawal of the machine. Then a round of holes can be drilled and the face fired. A drill, blast, excavate and support cycle is then adopted in the expectation that sooner or later weaker and less abrasive rock will be encountered, and the machine can perform satisfactorily, unaided.

BLASTING ACTION OF HIGH EXPLOSIVES

Most drilling and blasting operations in rock, other than tunnelling, involve the use of shot holes drilled parallel to one or more free faces (second free faces), to make best use of the reflected tensile stresses set up on detonation. The free faces also provide free passage for the shattered rock fragments produced by the explosive action in their flight away from the shothole. Delayed action techniques are essential to the process, enabling the use of available free faces whilst at the same time creating new free faces. These are subsequently utilised by the detonation of explosives in shot holes initiated later in the delay sequence. Breakage is caused predominately by tensile fracture, though the process can be much more complex. For breakage to occur by this method enough energy must be transmitted on detonation to account for losses in transmission through the rock. Fragmentation by tensile fracture may be the only means of breaking the hardest rocks with explosives, though fragmentation by crushing may be important in the softer and lower density rocks.

After the passage of the shock waves through the rock, the breakage process continues as the high pressure hot gasses produced by the detonation heave and churn the broken rock, projecting it from the shothole and placing it in a heap up against the newly formed face.

Unlike most other operators in rock, the tunneller looks at a fresh face of rock at the beginning of each cycle of operations. There is no second free face. He needs to create free faces parallel to the main access of the line of advance of the tunnel .The pattern of shot holes which he drills and the way in which he charges them, along with the sequence and the timing of the detonation system is the secret to successful tunnelling. Applying the "correct" or most appropriate cut to a tunnelling application, ensures that the planned regular advances produced by each cycle of operations (that is the "pull") approaches 100% of the depth of the drill holes in the face.

THE DESIGN OF THE CUT

The cut may be produced in a number of ways. If short advances are required, the cut may be formed by drilling a pattern of angled holes so that at the face they are well separated whilst at the back they converge almost to touch. A number of rows of holes are usually drilled and the rows may be either in the vertical or horizontal plane. Usually, the horizontal is chosen since this facilitates drilling. With this type of cut, the depth of pull is limited by the width if the tunnel This is particularly the case in small or narrow tunnels, and since drilling the angled holes requires that the driller and his drill lie across the line of advance of the face, only a limited member of drilling machines can be employed. Wedge cuts are popular with miners with a great deal of experience in the strata conditions of a particular mine. Where computer controlled rigs are employed on large faces this method often gives the best economy of explosive consumption and minimises the drilling. In any wedge cut there is the possibility that a large piece of rock from the undrilled portion of the wedge will be projected forward with potential for damaging machinery and dislodging supports. The chances of this occurring can be minimised by incorporating into the round a lightly charged stab hole to break up any such lumps.

Where the emphasis is on maximum advance, long pulls and the ability to operate a number of drilling machines at the face, parallel hole cuts (usually called "burn" cuts) have been developed to a degree where advances of several metres, unheard of at one time, are now readily achievable.

Parallel hole cuts incorporate within the pattern uncharged or void holes, which if sufficient in number, diameter and location are the key to the success of the round of shots. Whilst the wedge cut throws out a wedge of rock, the burn cut breaks out a cylinder of rock on a line parallel to the axis of the tunnel and then ejects the material forward through the void holes. The only second free faces here are the surfaces of these uncharged holes. Detonation breaks the rock around these voids systematically using delay firing at a rate which will enable the debris to fly through the void without choking it. It is the controlled use of delayed action shotfiring techniques coupled with a prudent choice of explosive characteristics which will make this possible. Overcharging causes excessive breakage and bulking of the debris to too great an extent. Too fast a sequence of detonations results in choking the void with debris. Where the cut has been overcharged or the detonation sequence too fast, the sockets contain powdered rock which has been compacted so that the holes later in the round have been unable to unload their debris forward. When the cut is undercharged and the whole round of shots fails to break to the back of the holes, cracked and shattered rock can be seen in the sockets of the cut holes. An experienced tunneller can recognise cut "freezing" and reduce his charges. If he sees signs of undercharging he can likewise remedy the situation by increasing the charges or the number of charged holes.

Void holes in the cut may be the same diameter as the shot holes in the round or they may have a much larger diameter. When hand held machines are used, it is usual to use only shothole sized holes. Where drilling rigs are being used it is not an unacceptable chore to drill one or more void holes with a larger diameter. A number of patterns of shot holes are popular with different shotfirers. The ones with a large number of large void holes perform the best, those with void holes located between the charged holes assist with the prevention of desensitisation and sympathetic detonation of the explosive.

Explosives choice for the cut must be able to accommodate the requirements outlined. An explosive with a high density and high bulk strength could well be concentrated at the back of the cut holes, causing cut freezing at depth and undercharging at the collar. Additional explosive charged at the collar can cause the rock at the collar to be ejected violently, but have no effect on the performance of the back of the cut.

The tunneller needs to charge the cut with a lower density explosive, so that each shot hole can be loaded to within a few hundred millimetres of the collar without producing too violent an effect and at the same time unloading the void of rock throughout its length.

THE DRILLING PATTERN

The spacing and burden of the shot holes around, beneath and above the cut will be decided on by experience based on the hardness of the rock, its strength characteristics and its structure in terms of discontinuities and variations in quality. Allowance will need to be made for the deviation which can be expected with the type of drilling being used. The maximum distance between holes will be decided by reference to complex formulae or more often by rule of thumb. In general, holes below the cut will have less burden than those to the side of the cut or above the cut.

The need to charge the full length of the shot hole is less critical in the bulk holes around the cut. Nevertheless, fragmentation is always an important consideration. Shot holes which are less than about two thirds full of explosive can be said to be under-utilised to some extent. The stronger the explosive, the more energy can be concentrated in each hole and the wider the hole spacings.

The location of trimming holes is considered separately.

CHOICE OF EXPLOSIVE

The characteristics of an explosive suitable for tunnelling include its strength and its velocity of detonation, both of which influence its fragmenting ability. Density is an important consideration. Tunnellers prefer low density explosives, since they like to spread their explosive along as much of each shothole as possible, without overcharging. Choosing an explosive of the correct density achieves some economy, and it also assists in producing optimum rates of advance, good fragmentation and an acceptable muck pile profile. A most important feature is the explosive's fume characteristics. In addition to the production of gases which will not support life, the oxides of nitrogen and carbon monoxide can be produced by some explosives. Compositions which produce more than acceptable quantities of these gases are unacceptable for tunnelling. A high oxygen balance leads to the production of nitric oxide and nitrogen dioxide in the explosion fumes, which are, therefore, highly toxic. A low oxygen balance generally gives higher explosive power, but produces poisonous carbon monoxide. It is the intimacy of mixing of the ingredients which influences the overall fume characteristic of an explosive. Oxygen balanced explosive products with coarse ingredients produce higher quantities of noxious gases than products which are more intimately mixed. Poor water resistance is always a consideration where conditions are not totally dry, since explosives which are not resistant to water will often form large quantities of noxious gases when fired wet. Re-entry times to the tunnel face are critical to the overall rate of advance. Slurry and emulsion products have shown up to a tenfold reduction in the quantity of nitrogen dioxide and a two fold reduction in the quantity of carbon monoxide produced when they have replaced nitroglycerine based explosives.

DESENSITISATION OF EXPLOSIVES

Desensitisation of explosive can occur in certain circumstances. Desensitisation in this context refers to the loss of sensitivity (the ability of the explosive in the form of a column, to maintain a steady velocity of detonation along its length). Desensitisation of explosives can occur when an explosive which is not completely waterproof is used in wet conditions and the composition becomes waterlogged. In this condition the explosive may detonate with less efficiency than normal, or the reaction may be incomplete. The temperature of the reaction will certainly be lower than the designed temperature. The result will be poor performance and the production of noxious gases which will seriously increase the safe re-entry time. If the explosive becomes too wet all or part of it may fail to detonate. Desensitisation occurs most often as a result of inadvertently increasing the density of the explosive above the limit at which it can continue to detonate. Channel effect, which causes the detonation of a column of explosives to fade out, occurs when detonation along the perimeter of a cartridge of explosive advances some way beyond the detonation front. When this occurs the unfired part of the charge increases in density. When this density increase exceeds the density above which the product is sensitive at that diameter, the detonation process ceases. If the hole diameter is much greater than the cartridge diameter the effect can be very serious. Channel effect most commonly occurs with explosives in rigid cartridges.

The least understood cause of desensitisation of explosives in tunnelling operations is the effect of detonation in one shot hole on the explosive in adjacent holes which are due to fire later in the delay sequence. The effect can be quite difficult to diagnose precisely, and produces sockets at the hole ends as well as undetonated explosive in the muckpile. This occurs in shotholes which, either by design or by accident are close to each other.

In hard rock, the dynamic shock wave through the rock which is the result of the detonation process in one shothole, may be passing through an adjacent shothole when that shothole detonates. The pressure of the shock wave from the first hole can distort the second shothole and increase the density of the explosive to such an extent that it will not detonate. If the explosive is sensitised by micro balloons they may shatter, if by sensitising gas bubbles, they are squeezed to an extent where sensitivity is affected. The time taken for the shock wave to travel such short distances as are involved, say 200 to 300 mm. in and around the cut, is of the order of 0.1 to 0.2 ms, so dynamic desensitisation of this type occurs in shot holes containing detonators of the same delay number and as a result of scatter on the nominal timings of the detonators. Later delays can be affected. The shot holes recover their shape as soon as the shock wave passes, and often the explosive recovers within such a short space of time that it performs successfully when fired by a detonator of later period delay.

In softer rocks it is possible for the detonation of one shot hole to distort the rock in the vicinity of adjacent unfired shot holes. These shot holes do not recover their shape, and the explosive within them is under static pressure when they in turn are detonated by a detonator of a later delay period. In fractured rock, it is possible for gases from one detonation to increase the pressure in other shotholes, by the transmission of the products of detonation through the fissures. If the pressure increases to a high enough level desensitisation can occur by this process, the undetonated hole will be one fired on a later period delay. The frequency of occurrence of these three events is probably in the same order and of the same significance as they have been described.

Nitroglycerine gelatines have often been found to be too strong and dense. Cautious miners, also interested in economy of explosives consumption (because the cost of explosives was extracted from their wages), adopted a practice of inserting short sticks of wood between each cartridge to spread the explosive's energy throughout the cut. Nitroglycerine powders with their lower densities and bulk strength have been found to be more suitable in this respect. The need to protect them from the ingress of water is a problem. Slurries and emulsions with their lower densities have much to offer in this respect.

Different types of explosive react differently to these effects. Sensitive nitro-glycerine explosives are least affected by desensitisation. ANFO is compressed to a level where the prills are crushed and the composition becomes dead pressed and will fail to detonate. Slurries, however, are resilient and tend to recover in time. Complete recovery takes up to

100 ms. Specially formulated compositions are necessary if the problem is to be eliminated. Emulsions sensitised with micro balloons are prone to dynamic shock failure.

Dynamic desensitisation can be overcome by incorporating void drill holes in the pattern. Static desensitisation can best be avoided by ensuring that the drilling and delay pattern has sufficient opportunity for free faces to be formed. Desensitisation by pressure from escaping gases can only be solved by the correct choice of explosive.

SYMPATHETIC DETONATION

Sometimes shotholes are placed so close together that the intense shock wave from one detonation initiates and detonates the explosive in adjacent shot holes. Some nitroglycerine explosives are prone to this effect. It is made worse by the presence of water. If it appears in the cut area of the design, it can cause cut freezing. Sympathetic detonation in the cut area can be reduced if not completely eliminated by drilling void holes between the charged holes. When sympathetic detonation does occur it can cause a large increase in the ground vibration generated by the explosions in the cut.

SMOOTH BLASTING

In tunnelling operations, the competence of the roof and sides to offer support to the excavation needs to be maximised. The shape of the excavation and the degree of overbreak affects the mucking operation, the support requirements and even the efficiency of the ventilation.

Whilst there has been a general decline in the use of standard strength detonating cord, the range of detonating cords available has increased and a variety of new applications have been found.

The earliest smooth wall blast design surfaces were achieved using small diameter nitroglycerine gelatine cartridges strapped at intervals to standard detonating cord, which were then located in the perimeter holes fastened to canes. This method produced the satisfactory de-coupling effect. The method was tedious, though the results were often encouragingly effective. This method has been much improved by the introduction of a coupling clip which holds detonating cord and cartridges together spaced to suit circumstances. In another method which ensures the accurate placing of the charges along the line of detonating cord and facilitates charging, the nitroglycerine gelatine explosive was provided in pre-cut lengths of coupleable plastic tubes, the charge partly filling the tube. The length of the tube was chosen to achieve the required charge density.

Similar loading densities were provided by the use of low bulk strength nitroglycerine gelatine powder explosive cartridged in long and often very thin tubes. This method is available in the United Kingdom. Explosives manufacturers in Scandinavia have developed methods using explosives in small diameters, as low as 11 mm and have extended the principle to cover a whole range of activities. An extension of the principle to other explosive types such as slurries has found limited favour.

The benefits of using continuous charges led to experiments in the use of long lengths of flexible cartridges, mostly of slurries. Most of this activity took place in the USA. In the United Kingdom however, the use of high energy detonating cord has found favour. The use of a single line of detonating cord has proved effective. Cords with core loadings of 80 g/m or higher have been developed for this application. They are easy to use, and the charge concentration is constant. These are often prepared so that they can be charged directly after the primer cartridge into the perimeter holes in a tunnel round, without the need for cutting at the tunnel face. The excellent results produced help to give confidence that the roof and sides are stable and the surface exposed makes setting supports easier and adds to their competence.

BULK EMULSIONS

Site mixing ANFO has not found much favour in construction tunnelling in the United Kingdom construction industry. In mining it is much more commonly found. In Great Britain, mixing ANFO is carried out under the Section 50 of the Explosives Act 1875 and covered by Order in Council No 1485. Routine Licences cover the mixing of ANFO by hand at the blast site at quarries where the product is for immediate use. Where the product is to be used at a mine and, or is, to include other ingredients, a Non Routine Licence is required. The license states the permitted ingredients and outlines the conditions of use. Where it is intended to mix mechanically, the mixing equipment must be of a type approved by an Inspector of the Health and Safety Executive. Various types of mixer and mixer placer units are available.

Bulk emulsion/ANFO mixer trucks are in service supplying quarries around the country, with a range of bulk explosives whose range of characteristics is extensive. The high bulk loading densities and good degree of coupling with the rock ensure levels of efficiency which have enabled significant reductions in explosives cost to be made. The high speed operation of these systems coupled with the skill of the operators has had the overall effect of increasing blast size as well as cutting labour costs. The overall effect has been a marked reduction in the operating costs at some units.

Bulk emulsions can be manufactured, under Order in Council No 1485. The systems are well proven overseas. An unsensitised emulsion matrix is delivered to the site in bulk and is transferred to a charging vehicle. The vehicle travels to the blast site where the product is sensitised by the application of dosing chemicals and delivered into the shot holes. Powered delivery of the explosives and automatic hose handling, coupled with electronic logging of the data into a detailed blast record has been found to be valuable in monitoring performance.

An intimately mixed oxygen balanced explosive produces fumes which have only minute quantities of harmful gases. The inherent water resistance of the water in oil emulsion ensures that this is the case whether conditions are wet or dry. The energy content of the product can be modified to suit changing conditions. The process of sensitisation can also be used to accommodate the individual requirements of different parts of the round, by alterations to the density of the product. Thus, the cut can be charged with lower bulk strength explosives whilst the bulk holes can be charged more heavily. The sump holes at the bottom of the excavation can loaded more heavily. Very low density explosives for the smooth blasting of the sides and roof can be possible.

ELECTRIC DETONATORS

The increase in the use of electrical equipment caused concern during the 1960s about the risk of inadvertent initiation of standard electric detonators. Underground, low tension electric detonators had been used for many years in coal and other gassy mines. New ranges of detonators were being developed for the most hazardous situations and these are used extensively today.

In other mining applications, fuse firing was still common, although the method was being steadily replaced by electric shotfiring techniques using instantaneous, short delay and delay detonators. As this change was taking place, the introduction of ANFO, loaded pneumatically, was becoming more prevalent. Its use introduced problems associated with the generation of static electricity. More electrically operated plant was being introduced and the potential for premature initiation of electric detonators was increasing. Pumping water at tunnel faces or in sumps posed problems of earth leakage when electric pumps were used.

A number of methods of producing better quality initiation of modern explosives, minimising the environmental disturbance and the risk of premature initiation, were introduced around this time. The accuracy of delay times, the reduction of the incidence of misfires and the ability to give the optimum timings to avoid the excessive projection of rock from the blast were all considerations which led to the preference for using shock tube detonators, introduced at this time. Since then, these have been undergoing continuous development to refine these desirable features.

SHOCK TUBE DETONATORS

Nonel, the first shock tube system developed by Nitro Nobel of Sweden, was introduced to the market in 1973. It is one of the most important inventions in blasting technology this century.

The system is based on a plastic tube, the inside of which is coated with a reactive substance that maintains the propagation of a shock wave at a rate of around 2000 m/s. This shockwave has sufficient energy to initiate the delay element in a detonator. The reaction is contained in the tube and has no blasting effect, merely acting as a signal conductor.

Originally the tube was made of a carefully chosen plastic to an outside diameter of approximately 3mm. In the first instance the plastic was transparent. Heavy duty plastics and temperature resistant plastics were used in special circumstances.

The detonators are made with aluminium shells, containing a base charge, a priming charge, originally of primary explosive, and a delay element. Each detonator is sealed by crimping in a rubber sealing plug, which also protects a portion of the tube against wear. A length of Nonel tubing which fits through the plug is sealed at the free end.

The system contains a number of types and styles of connector unit. These act as relays. The shock wave from the Nonel tube is received, amplified and distributed to a number of receptor tubes. Plastic connector blocks hold a miniature detonator to amplify the shock. The connector blocks are designed to give mechanical protection to the transmitter caps, to slow down aluminium shrapnel from the caps and to bring any tubes into contact with caps. Transmitter caps are either instantaneous or have pyrotechnic delays. A connector specially designed to facilitate the coupling up at a tunnel face or in shaft sinking contains a double loop of detonating cord.

A range of detonators suitable for tunnel blasting was developed early in the 1980's. These had a delay of 100 ms between adjacent numbers at the beginning of the range, followed by detonators with a delay interval of 200 ms. At the high end of the range this delay increased to 500 ms. In all, 25 delay periods were - and still are - available. Detonation is achieved by connecting bunches of detonators together and binding them in detonating cord, usually as part of a bunch connector. When connection is complete, the network of connecting tubes is eased away from the tunnel face and the round of shots is fired using either an electric detonator as the initiator or a long length of Nonel tubing in the form of a connector.

The system can be used without risk of accidental initiation from any electrical source, and since the tubes will not initiate by shock, friction or heat, a charged face remains safe for work until the connectors are introduced. Charging can commence as soon as the drilling cycle has been completed, and pumping can continue until the time comes to withdraw workers for firing. The system is visually examined rather than checked as with electrical systems, and the bunch blasting system makes this relatively easy. Electric initiation of the round is by a simple series electric circuit consisting of just one detonator and a shotfiring cable. Where electrical hazards are still present a very high firing current detonator will usually overcome the problem.

NON PRIMARY EXPLOSIVES DETONATORS (NPED)

Most detonators contain both primary and secondary explosives. Primary explosives burn extremely quickly at atmospheric pressure, with burning transiting into detonation in a fraction of a millisecond. In a conventional detonator, the relatively slow burning in the delay element results in detonation of the base charge via the primary charge. In the new Nonel system, the same function is achieved by means of an initiation element. This gives increased safety and a significant reduction in the amount of lead in the blast fumes.

The detonator looks much the same as any other aluminium shell detonator. Internally, the sealing plug and tube arrangements are the same and the delay element is unchanged. The base charge is the secondary explosive hexogen. Only the initiation element consisting of a steel tube filled with a secondary explosive PETN is different. By having the secondary explosive in the initiation element divided into a number of charges of different qualities, the relatively slow burning in the delay element transits quickly to the detonation of the base charge.

The NPED detonator is much less sensitive to different stimuli than a conventional detonator. Higher safety levels can be achieved throughout its life. Manufacture, transportation, storage and handling are much less hazardous. On transport, for example, there is a reduced risk of mass explosion. This is best demonstrated by flashover tests. Resistance to mechanical impact has been tested by dropping a 5 kg steel ball onto detonators. The risk of premature initiation by impact is considerably reduced.

By the elimination of the primary explosive lead azide, the level of lead in a mine atmosphere is reduced.

NONEL TUBE

A recent development has been the introduction of 3 layer composite tubes. The layers consist of different grades of plastic with different properties. The innermost layer has good adhesion properties for the reactive substance. It also has good radial strength to prevent splitting when subjected to the stresses that arise when the shock wave travels through the tube. The plastic in the middle layer gives the tube its good tensile strength and acts as a barrier to oil and other chemicals. The outermost layer has good resistance to abrasion. It also serves as the cosmetic layer to which colouring of the tube is applied to aid recognition of the different ranges of detonators available. The method of construction of the composite tube reduce the 'memory' of the tubing and reduces its tendency to coil. Modern shock tube detonator assemblies, less prone to inadvertent initiation, with more reliable, robust tube with newly designed connector caps colour coded to make component recognition easier have added to the safety culture of today's mining and quarrying industry.

ELECTRONIC DETONATORS

The development of Electronic Delay Detonators (EDD) is just the latest stage in the refinement of detonator technology. By incorporating a microchip in each individual detonator, it will now be possible to program the delay of each detonator much more precisely than before. With conventional detonators the scatter of firing times around the nominal delay times must be less than 8 milliseconds to avoid overlap in the short delay range. In the tunnelling ranges of detonators, it is common to take advantage of the scatter times at the higher end of the series to fire off many more detonators than is consistent with the maximum instantaneous charge limits set by reference to a scaled distance/ m.i.c. graph. The electronic detonator's reliable accuracy of less than 1 ms reduces vibration and makes prediction much easier. This will bring immediate benefits in improved performance and vibration control.

EDD firing systems now being developed comprise up to 200 integrated electronic detonator (IED) firing caps and a programming console for the testing and programming of each IED. The delay is generated by an electronic circuit which is programmable from 1 to 6000 ms increments. The electronic section of the firing cap is contained in a slightly lengthened version of the current PVC plug. The IED is identical to an instantaneous detonator

The detonator circuit can be fired either manually or by computer. Using the manual system, each IED detonator is programmed individually using the programming console, which records their order, number and the delay. The detonators are then linked in parallel to the firing console, which checks the integrity of the circuitry and timing. The operator sets the parameters for each detonator. The computerised firing process is slightly different. Initially each detonator receives only an order number using the programming console. The pre-determined firing pattern is then transferred from a lap top computer assigning the pre-set delay to each IED before carrying out the integrity checks as described earlier.

In addition, there are a number of advantages to improved delay precision. The opportunity to download blast patterns from a computer is novel and allows the operator to re-program the blast sequence right up to firing time. The checking procedure has the potential for making this a very safe system. Blast initiation records can be stored for repeated use.

The degree of precision that can be achieved will provide tunnellers with greater control over fragmentation and vibration control, leading to a reduction in the environmental impact of blasts.

CONCLUSION

The combination of bulk emulsion technology and electronic detonators should ensure that the use of explosives in tunnels will continue well in to the future. It is to be noted by those developing products and systems that it is only their ability to respond to the circumstances of the time which will guarantee the use of explosives forever.

Definition and Control of Boundary High Walls at Butterwell Open Cast Coal Site Using Pre-split Rock Blasting

D. W. Horne

TAYLOR WOODROW CIVIL ENGINEERING LTD, MINING DIVISION, TAYWOOD HOUSE, SOUTHALL, MIDDLESEX, UK

1 INTRODUCTION

This paper reports mainly on work carried out during 1990 & 1991 at Butterwell open cast coal site, one of the largest sites ever operated in the UK, as part of the drive to maximise the quantity of coal recovered.

Operations began in 1976 at the northern end of the site with the development of the box-cut, and subsequently progressed steadily southwards. Early in 1990, excavation of the uppermost overburden bench had commenced in the final cut at the southern limit of excavation. It was necessary to consider the safety aspects of developing a 100-120m high final wall: as excavation of the lower levels progressed to the limits, parts of the high wall would have to stand for a period of eighteen months whilst coal was recovered, and until backfill was complete. The overall potential for catastrophic failure of the wall and any hazards which personnel might be at risk from whilst working at various levels below ground and close to the high wall, in a region where even minor rock fall or spalling might result in serious or fatal injury, had to be addressed. Additionally, the coaling limits set by the client had to be achieved and indeed the recovery of coal, thereby revenue, had to be maximised.

Traditional production blasting followed by excavation back to 'hard' ground, which tended to result in poorly defined,and in sections,unstable high walls, could subsequently lead to rock fall and associated hazards. In order to secure a safe working environment, this tendency required that relatively large safety benches be left at each high wall foot to accommodate any rock fall or spalling and prevent such material from falling into subsequent excavation levels. This arrangement would lead to the overall angle of the high wall, measured from the horizontal, becoming small.

This situation was not critical during the advancing mine state where high walls were temporary and indeed where safety benches had the added value of protecting the coal edge and preventing loss. However, at the limits of excavation, the high walls had to stand for a considerable period of time and large safety benches would lead to coal, which might otherwise be recovered, being left in-situ.

A decision was therefore made to do the following:-

i) Carry out an investigation to ascertain what the steepest overall high wall angle could be, bearing in mind stability considerations and suitable factors of safety against catastrophic failure.

ii) Define the limits of excavation by pre-split blasting.

iii) Reduce the number of safety benches given the safer, cleaner wall which would be produced by using pre-splitting.

To satisfy the first requirements, professional geotechnical engineers were engaged to carry out a site inspection and subsequently report with safety factors for a variety of worst case scenarios taking into account high ground water tables, adverse dip direction, joint patterns and the rock/coal/seat earth discontinuities. The study confirmed that factors of safety were acceptable for a high wall of 50° to the horizontal at the Butterwell southern boundary.

To effect the pre-splitting, a system was developed which is a derivative of, but noticeably different from pre-splitting in the traditional sense.

To strike a balance between a hypothetical extreme of a single vertical wall with no intermediate bench, and the traditional bench at every coal seam, a series of 12-15m high benches (each through 1, 2 or 3 coal seams) was chosen.

2 PRE-SPLIT ROCK BLASTING

2.1 Introduction

Much research work has been carried out to define why and how pre-splitting works: the physics and mechanics of the process are complex, and this paper does not attempt to emulate the work of others who have published learned papers on the subject.

Nonetheless, the essence of the traditional technique is explained so that the difference between tradition and the modified method will be clear.

2.2 Traditional Method & Theory

Closely spaced holes are drilled along the line to be pre-split, the holes are loaded with de-coupled charges, and all charges are detonated simultaneously. On initiation a compressive shock wave, generated at each hole, travels out radially from each hole. The shock waves first meet at points midway between holes, and are reflected as tension waves. Provided the tensile stress generated is in excess of the rock tensile strength, cracks develop. A crack is therefore created linking all the holes in the pre-split line and the high pressure gases developed in the boreholes further 'wedge' and increase the crack.

The distance between adjacent holes is normally set in the range 8 to 12 borehole diameters, and hole diameters have generally been in the range of 50-100mm. Charging rates vary depending on hole diameter, rock type, and expert opinion, but most obviously on results. Charges used are either pipe charges of a diameter suitable to yield a complete column at the required loading rate, or cartridges spaced to achieve the required rate. Charges are generally of much smaller diameter than the borehole and are said to be decoupled. This is very different from the coupling achieved when bulk blasting with loose poured explosives. Detonating cord is used to initiate the explosives over the full depth of the hole, the detonating cord itself being initiated at the surface.

2.3 Modified Method

The major factors which prompted a search for and development of a modified pre-split system, were the site blasting restrictions. Whilst it is known that vibrations arising from the detonation of de-coupled charges are very much lower than for coupled charges of a similar mass, blanket blasting restrictions (based on the Morris formula, with other arbitrary reductions in permissible charges) obtained which (a) limited the maximum charge per delay (severely at the site limits) and (b) required that only one hole per delay may be fired.

It was therefore necessary to devise a system whereby -

a) every hole fired at a discrete time;
b) the time between adjacent holes was very short to help avoid the possibility of cut-offs;
c) use was made of the available but relatively large diameter drilling equipment, and additional specialist equipment was not required;
d) full advantage was made of the relationship between borehole diameter and spacing to minimise drilling costs;
e) an effective pre-split was generated!

A system was devised using short delay electric detonators together with a sequential blasting machine which satisfied (a) & (b), used 159mm diameter boreholes (c) at 2.0 metre spacings (d), yielding (e).

Figs. 1, 2 & 3 'Blast Specification for Pre-Splitting to Butterwell High walls' show how the holes were charged and the delay arrangement and connections used to achieve the very short nominal delays between adjacent holes, and the typical bench arrangement with hole depths.

The theory as to why this method still produces a pre-split is quite different from the traditional mechanism and is a subject which might form a separate paper. In essence, the compressive shock wave from charges detonating in one borehole travels way beyond the next adjacent hole before the second hole detonates, and so there is no mid-point meeting of shock waves from simultaneous detonations in adjacent boreholes. The most likely mechanism is therefore the reflection of the shock wave from one borehole at the free face provided by the next borehole, and crack propagation back from that free face. The fact that the charges were de-coupled doubtless afforded protection from any possibility of pressure desensitisation: no failed charges were discovered in more than 10,000 pre-split holes used. The nominal delay of 10 ms between adjacent holes was probably much too short a time in which movement of rock across into adjacent holes could occur and cause cut-offs. Again no cut-offs were discovered in any of the pre-split work.

Purely vertical drilling was employed given the simplicity of control this provides, especially with large diameter rotary drilling with stiff drill tubes. The option of angle drilling was considered but it was decided the additional difficulties of controlling azimuth bearings would far outweigh the advantages of angle drilling. The main advantage of angle drilling in relatively flat strata to produce a sloping rock face, is the improved stability arising from the reduced propensity for toppling failure. However, any loose material falling from a sloping face will also gain a component of horizontal velocity and therefore roll and bounce, conceivably across intermediate safety benches. On the other hand, any loose material falling from a vertical face does not gain other than a vertical velocity which is readily halted at a safety bench. Given that the safety benches were introduced almost exclusively at coal seams, the seat earth provided a natural adhesive to aid capture of any spall at the bench. Reference to Figure 4 will assist in the understanding of these important points. The drillers were congratulated for their simple practical ideas to ensure that holes were drilled on line, repeatedly at the required spacings, and worth describing here. To ensure a straight line of pre-split holes, a simple pair of 'gun-sights' on the rig over which the driller could look to a pair of back markers were developed. To ensure accurate spacing, on completion of the first hole in a line, an 'anchor' such as a large heavy steel nut, was tied to the rig and laid on the ground at spacing distance from the first hole. The rig moved off towards the second hole, and was clearly in exactly the right location when the 'anchor' just fell into the first hole. The pre-split line can only be as good as the drilling, and care in accurate collaring, spacing and verticality contributed greatly to the success of the technique. The fact that the highwall was visible to all was a considerable incentive: the drillers' work was on display, they had to get it right!

A range of experiments with charging rates and indeed with hole spacings was carried out, before finding an average of about 1.0 kg/m @ 2m c/c ideal in the sandstone and shale benches. Referring again to Figure 1, it can be seen that a variety of explosive types in differing configurations and distributions were used. There was no apparent or noticeable difference resulting from either type or distribution of explosive over the ranges used. Within the range of particular conditions and parameters used, the required charging rate is proportional to a function of the spacing distance. Termination of the pre-split holes in coal seam was also a careful design feature: the softer nature of the coal avoided the need for a heavier toe charge, and provided a natural break point for the split.

Once the pre-split blasting had been carried out it was necessary to take care with the location of the bulk blast holes closest to the pre-split line and it was found that drilling bulk blast holes about half the normal burden away from the pre-split line yielded well broken rock up to, but not across the pre-split line. The firing sequence of the bulk blast adjacent to the pre-split was also apparently significant; a cleaner wall tended to result when the "firing direction" was along the bench rather than across the bench, and is thought to be related to the shearing action along the wall which certain bulk blast firing orders can generate.

It was also found that during bulk excavation exposing the pre-split, the wall should be carefully 'scaled' by machine to remove any loose rock, which might otherwise spall off during subsequent coaling operations adjacent to the wall thereby compromising the safety of coaling personnel.

The results were very safe, clean, well defined walls which could be readily inspected, and in which any movement was obvious, and 92,000 tonnes of additional coal which more than financed the extra cost of pre-splitting. A safe and cost effective technique.

The success of this exercise demonstrated that in certain circumstances it was feasible to safely approach far closer to the site boundary with the excavations and therefore increase the recoverable tonnage from the site. As a direct result it was possible to define a further extension beyond the original limit of excavation using the same system, and by steepening the overall highwall angle to 60° still with acceptable safety factors, a further 193,000 tonnes of coal were recovered from the site. Such a steep highwall angle only became possible as a result of the extensive geotechnical investigation, and the intimate site knowledge.

Fig 4 is a drawing illustrating the highwall profile typically achieved in the past by excavating back to hard, the profile actually achieved by employing the pre-splitting system developed, the reduced number of safety benches, and the extra coal.

Overall, it was satisfying to have developed a control system for high walls on an open cast site which produced a range of benefits. These included greatly reduced survey and setting out work; precise control by way of obvious excavation limits up to which prime movers could accurately work; easily achievable design coal limits; clean, safe, stable walls and benches which were readily inspected; and the opportunity to increase the amount of coal which could be won from a given site. The system was also self financing, the extra coal paid for the pre-splitting costs.

So it is fair to say this method of work achieved enhanced definition and control of high walls, and did so by making good use of 'explosives in the service of man', just a small part of the 'Nobel Heritage'.

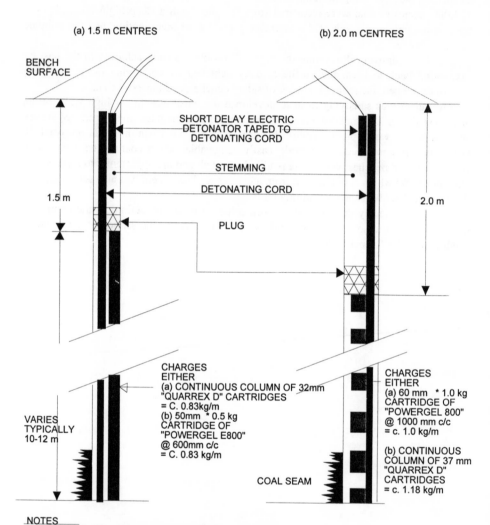

NOTES
1. ALL HOLES DRILLED VERTICALLY, USING 159 mm BITS.
2. STEMMING LENGTH IS INCREASED TO AVOID EXCESSIVE
SHATTER WHEN WORKING IN WEATHERED ROCK HEAD.

Figure 1   *Blast specification*
Pre-splitting to Butterwell highwalls
Hole loading diagrams

PLAN	DELAY No	NOMINAL DELAY(ms)	CHANNEL No	CHANNEL DELAY(ms)	NOMINAL FIRING TIME(ms)	INITIATION SEQUENCE
	12	305	1	0	305	1
	12	305	2	10	315	2
	12	305	3	20	325	3
	13	335	1	0	335	4
	13	335	2	10	345	5
	13	335	3	20	355	6
	14	365	1	0	365	7
	14	365	2	10	375	8
	14	365	3	20	385	9

PREVIOUS PRE-SPLIT HIGHWALL

NEW PRE-SPLIT LINE

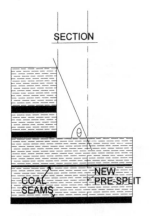

SECTION

NEW PRE-SPLIT

COAL SEAMS

NOTES

1. BLAST FIRED USING THE REO BM175 -10PT SEQUENTIAL BLASTING MACHINE WITH THE INTER CHANNEL DELAYS ALL SET TO 10 ms.

2. NEW PRE-SPLIT LINE IS EASILY SET OUT AT A FIXED OFFSET FROM THE PREVIOUS HIGH WALL SO AS TO ACHIEVE THE DESIRED OVERALL HIGHWALL ANGLE θ

3. HOLE DEPTHS VARY ACCORDING TO THE REQUIRED BENCH HEIGHT & COAL SEAM HORIZONS.

Figure 2   *Blast specification*
Pre-splitting to Butterwell highwalls
Initiation sequence, layout and geometry

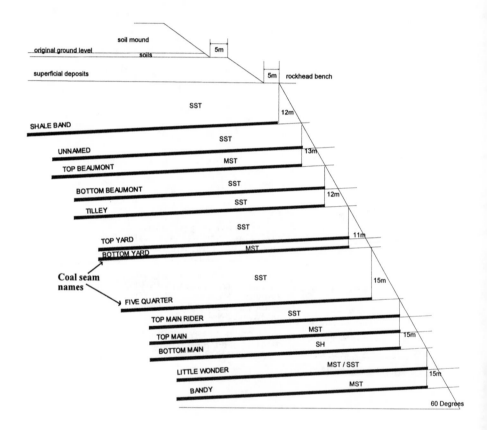

Figure 3   *Blast specification*
Pre-splitting to Butterwell highwalls
Generalised cross-section showing pre-split faces

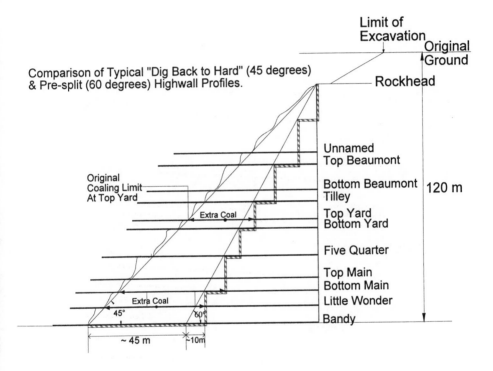

Figure 4 *Definition and control of boundary highwalls at Butterwell*

Evening Lecture

Alfred Nobel. The Ultimate Rocket and Constructive Explosions in Space

A. Hansson

REACTION ENGINES LTD, 20 LEYBORNE AVENUE, LONDON W13 9RB, UK

1 INTRODUCTION

Even if gunpowder has been used since the Song Dynasty in China (960-1279 AD) and later around the World, explosives are mainly today associated with Alfred Nobel.

There are essentially three different space propulsion systems:

- the rocket principle, where the vehicle carries its propellant material and energy source;

- a system in which energy is supplied from an external source like nuclear explosions;

- and a system which impulse (and energy) are applied by an external agency as in solar sailing.

There is not any system, even in concept, that can match the specific power of the chemical rocket engine.

The problem with this propulsion system is that it is expendable and that it cannot be safely aborted in a way that saves its cargo. Ten years ago, in 1986, in the period from Jan 28 to May 30, the following launchers failed: Challenger STS (28 Jan), Titan-34D (18 April), Delta-3914 (3 May) and Ariane-2 (30 May), indicating that satellite launching with present technology is a very uncertain transport business.

The present operational space launcher system still suffers from this disadvantage and leads to the conclusion that cargo carrying must wait for a vehicle to be constructed, a feature that is unique in any transportation system.

2 ALFRED NOBEL

Alfred Nobel first came into contact with aerospace in 1886 as a financier (among others) behind the tragic expedition to the north pole in a controllable balloon, by a fellow Swede called S.A. Andre. A vivid description of this expedition is given in *The Flight of the Eagle,* by Per Olof Sundman (Pantheon Books, 1970) . However, his interest was not limited to this particular venture but extended to remote sensing, both from balloons and other devices which have been named 'air torpedoes'. Four months before he died in 1896, he wrote:

> "I suggest sending up a small balloon with a parachute and camera
> together with a small timer and ignitor. At an optimal height the
> parachute is automatically separated by the operation of the timed
> ignitor and returns to the earth with the pictures in the camera".

Already during the 1880's, Nobel had been interested in a rocket propellant for assistance in artillery. The reason was that dynamite had been tried but could not be used because it exploded under the high acceleration forces involved with a shell in trajectory. In the USA, several million $ had been spent on 'dynamite artillery' but explosions almost invariably took place in the barrel.

A Swedish officer, called Wilhelm Teodor Unge, contacted Nobel with an idea of using rockets fired at the highest point in an artillery shell's trajectory, in order to extend their range. Alfred Nobel became very interested in this idea since it had the advantage of slow acceleration in rocket assist. Unge left the military and set up a firm named Mars Ltd. (A.B. Mars, in Swedish) and obtained a suitable propulsion fuel from Nobel in the form of the high performance Nitroglycerine/Nitrocotton propellant "ballistit". The first launch took place on the 18 of September 1893 and at the time of Alfred Nobel's death , Unge had solved not only the stability problem that had forced rockets out of artillery, but also he had extended their range. The solution was to incline the propulsion so that the rocket rotated but remained on its own centreline.

In 1892, Alfred Nobel concluded:

> "I confess that flying inspires me, but we will not be able to solve
> its problems with balloons. When a bird has reached such a high
> speed that it can overcome gravity, it can then fly with only the
> minimal use of its wings. This is not done by magic. What a bird
> can do, so can humans. We must have airfleets propelled at high
> speed. A dove can fly between Paris and San Remo in 3 hours..."

With Nobel's death in 1896, Unge and AB Mars got into financial problems since the Nobel prize trust then took financial precedence in the Nobel estate provisions.

From comments made by Alfred Nobel, it is obvious that he regarded speed as essential, a view expressed in the comment:

" A thought, via electricity, could travel round the Earth in about a quarter of a second".

Alfred Nobel thus rightly viewed the future of air travel in the form of accelerating vehicles and not in the present cruising type of vehicles.

It was not until 1901 that it was possible for Wilhelm Unge to reconstruct AB Mars. In 1908, after renewed success, Krupp took control of the patents, partly with a view of blocking a competitor in canon-based artillery.

Wilhelm Unge died in 1915 and a few years later Krupp publicly announced that they were no longer interested in further work on rockets.

After the war, Krupp's head of ballistic research, Otto von Eberhard, returned to Unge's legacy and, according to Willy Leys books,'ballistit'was, once again, used   by the government rocket establishments.

3 THE ULTIMATE ROCKET

Wilhelm Unge's life partly overlapped with that of the great theoretician Konstantin Tsilkowsky. In 1903, the year of Wilbur and Orville Wright's 'Flyer' at Kitty Hawk, Tsilkowsky proved not only that it was possible to leave Earth, but also that the most economical fuel was liquid hydrogen and liquid oxygen. Hermann Orbeth's *The Rocket into Interplanetary Space* was published in 1923 and in 1926, Robert Goddard's first rocket travelled 55 m.

Meanwhile, the aeroplane was poised to overtake sea transport as the principal transatlantic passenger service. This was the culmination of 50 years of development which started in the 1920's and resulted with a number of airlines coming into operation round the world, mainly financed by mail and high specific value goods. Already in 1919, 5000 passengers were carried by air. The limiting factor was that the equipment, being either ex-military or based on military technology, did not make for economic payloads and was not very reliable with the resultant high maintenance costs.

In 1934 Douglas introduced the DC-2 as a precursor to the revolutionary DC-3. The DC-3 was a commercial airliner, reliable, and based on the existing engineering. With the DC-3, operators made a profit carrying passengers , following Henry Ford 's conclusion a decade before, that passengers were the key to the development of air transport.

Progress continued and in 1939 western airlines carried 2.5 million passengers even without any of the airport facilities which are now regarded as indispensable. Airports started to appear in the 1940's and by 1945 passenger traffic reached 9 million , numbers which increased to 300 million per annum in the 1970's when air cargo began to grow rapidly with the introduction of high capacity aircraft.

By the end of the 1950's, technology was sufficiently advanced for the first orbital launch, in the form of Sputnik I, which took place in 1957, and was followed by the first human in space, namely, Yuri Gagarin in 1961. A listing of the first satellites is given in table 1.

TABLE 1 LIST OF FIRST SATELLITES

Country	First Satellite	Launch Date	Payload Kg.	Failures
USSR	Sputnik 1	1957 Oct 4	83.6	0
USA	Explorer 1	1958 Feb 1	14.0	1
France	Asterix 1	1965 Nov 26	40.0	0
Japan	Ohsumi	1970 Feb 11	38.0	4
China	Dong Fang Hong	1970 Apr 24	173.0	
U.K.	Prospero	1971 Oct 28	66.0	1
Europa	-	-	-	3*
Ariane/ESA	CAT 1	1979 Dec 24	1602**	0
India	Rohini 1B	1980 Jul 18	35.0	1
Israel	Offeq 1	1988 Sep 19	156.0	0

NOTES: Europa (marked *) was developed by European launch vehicle development organisation but the programme was cancelled after the third failure to achieve orbit. CAT 1 (payload marked **) was the first test flight of the Ariane 1 launch vehicle developed by ESA and then marketed by Arianespace: the payload mass included 1385 kg of ballast. All of the launch dates are based upon GMT, thus ignoring the local date and time that the launch took place (e.g. Explorer 1 was launched on 31 January local time).

By this time, the USA Apollo programme and its technological and operational spin-offs, had fulfilled many of the initial ambitions of the pioneers of the space age, with Earth satellites, people on the Moon and in Earth orbital stations. But two important differences existed: the scale should have been larger and the transportation should have been a synthesis of aviation and rocketry. Even today, each space mission is a hand-crafted entity, with assembly and quality testing at the launch site, where as many of the parameters as possible, are monitored. And, since 2.5% will go wrong during take off, it is necessary to be able to destruct the vehicle instantly. The continuing acceptance of this sort of situation, as we approach the millennium does no credit to the science and practice of astronautics.

How can this situation be changed?

As much as 90% of the weight of the Saturn rocket, used in the Apollo programme, was fuel, stored in its first three stages and jettisoned before the rocket left Earth orbit. Thus, a return to Apollo-like engineering is not the answer. There are at present two modern myths. The first states that the use of solid or low-technology liquid propulsion expendable launchers can achieve very low launch costs. The truth is that any launch vehicle designed to meet the required level of reliability will have a very expensive development programme, irrespective of its technology. This, in combination with an increased size due to low technology obliterates any the potential gain.

The second myth states that only with the use of very advanced materials, exotic propulsion and hypersonic aerodynamics can a vehicle fly into orbit as a single stage vehicle. The truth is that hybrid engines, together with the available materials and a systems engineering approach to the problem, yields a viable solution. This is the ultimate rocket - a rocket hiding inside an aircraft - and we call it:

SKYLON (see figure 1).

SKYLON could do for space what DC-3 did for aeronautics. SKYLON is designed to accept 4.6 m diameter payloads, standard cargo containers and, in its second generation, it can carry 60 passengers. It can be fuelled by liquid oxygen and hydrogen under no vent conditions at the runway, with a 2 hour hold before propellants reprocessing. In short, it can be introduced from existing launch sites. Later, as happened in aeronautics, dedicated spaceports, on the equator, will reduce costs to handle cargo.

As every terrestrial transport depends on fuel logistics, so will it be for space and, if economic transport can be established, the main cargo to orbit in the mid 21st century will be propellants.

With the rate of progress over the past 100 years, it is possible that by 2069, 100 years after the first Apollo Moon landing, we could have substantial industrial sites on the Moon to support Earth's population without damaging the Earth's biosphere any more than we already have. Lunar liquid oxygen would be used, but hydrogen is still likely to have come from Earth. The equatorial spaceports could each handle 140 flights per day, with perhaps approximately one million people per year in each direction. Cargo could reach 400,000 tonnes per year, half of which would be to the Moon and the other half for deep space missions. The trade flow to Earth would have increased up to a billion tonnes a year, including refined metals and finished industrial products. In fact, only the industrialisation of space resources will be able to meet the already increasing demands from the 75% of the Earth's population who are in the process of acquiring the same level of technological standards as are currently enjoyed by the privileged 25%, while still leaving the Earth fit for inhabitation.

The fact that the engineering of a space plane is more demanding than a present aircraft does not mean that its operation is more complex, or that we should not introduce it out of fear of its possible military potential. Its potential to form a leap into a sustainable future for the Earth's population is far greater. The only thing that is lacking now is the political will and assent for such accelerating space transport.

It is worth remembering Tsiolkowsky's words from 1929:

> "Most people consider astronautics a heretical idea and refuse to entertain it at all. Others are sceptical, regarding it as an absolute impossibility, while others are too credulous, considering it a simple matter, easily accomplished. But the first inevitable failures

will discourage and repel the fainthearted and destroy the confidence of the public".

4 CONSTRUCTIVE EXPLOSIONS IN SPACE

We are now at a transitional stage of discouragement. With a new view of astronautics, however, we can be on our way to the stars once more. To leave the Earth we will need space planes for the stars, and some form of nuclear pulse propulsion. The energy needed to ignite a small fusion explosion around 10 MJ is contained in some 1 kg of high explosive. The problem is to concentrate this energy in a volume less than 1 cubic cm and release it in a time shorter than 10^{-8} s.

In 1891, Herman Ganswindt proposed to propel rockets by a series of controlled chemical explosions on the same basis as the repetitive fuel explosion principle of the internal combustion engine. In 1955, Stanislaw Ulam and Cornelius Everett proposed to use nuclear fission bombs instead of controlled chemical explosions. This was the Project Orion which had a budget of $ 10 million before its cancellation in 1965. Before then, in 1959, a 110 kg prototype, Hod Rod, with chemical explosives, had proved the principle. Next followed the British Interplanetary Society's Project Daedalus which ran between 1973 and 1978. Here, fusion was selected to send a probe to Barnard's star. The fuel consisted of pellets of deuterium-helium fired at 250 per second. Fusion explosions can be made smaller than fission explosions because their products are much shorter in range than fast neutrons in fission.

There is some hope based on compact glass lasers but the low efficiency and hence large size, makes such a system unrealistic for space propulsion. A different approach was made by Winterberg. He suggested that a compact chemical explosive laser should be used to directly ignite a thermonuclear micro-explosion. The critical element is to make this high explosive transparent so that a cylindrical laser pulse can trigger a photon avalanche in the population inverted gas column of the burnt high explosive (see figure 2.)

Alfred Nobel with his insatiable intellectual curiosity and his genius for finding solutions to problems regarded by others as without solution would have found the challenge irresistible. He had already in the 1880's realised that the fundamental and key requirement was higher and higher speeds. Interstellar transport would be a fitting recognition and tribute to his concept of:

"Round the Earth in 1/4 of a second"

And a tribute to the excellence of achievement in the advancement of science and knowledge to which Alfred Nobel was so devoted and which he honours in the prizes which carry his name.

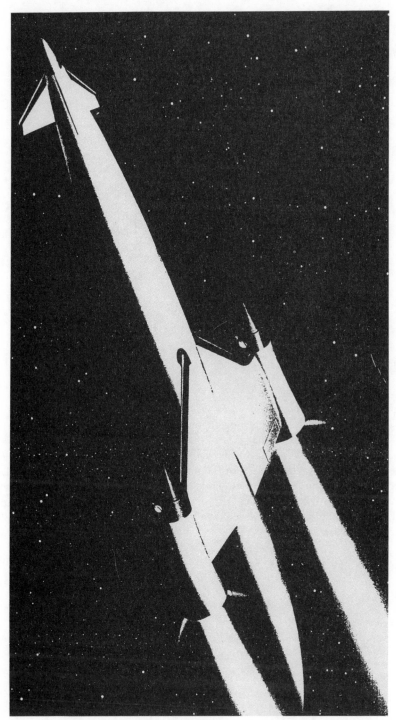

Figure 1: Artist's impression of SKYLON

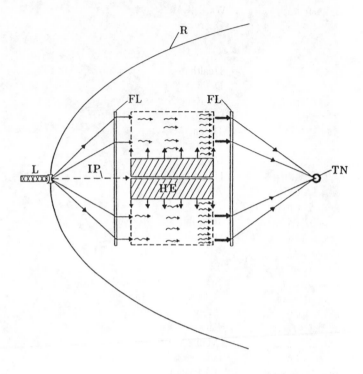

Figure 2: Chemical explosive laser for thermonuclear microexplosion ignition. L is the trigger laser, HE is the high explosive, TN is the thermonuclear target, R is the reflector for transmitting thrust, FL are the Fresnel lenses and IP if the ignition pulse (adapted from F. Winterberg' article at *Missions to the Outer Solar System and beyond*, Turin, June 25-27, 1996).

List of Participants

Anderson, Mr John	Wardell Armstrong
Aspin, Mr Robert	Quarry Services Ltd Ireland
Bamfield, Dr Peter	RSC Industrial Affairs Division
Barr, Dr Hugh	Health & Safety Executive
Baker , Mr Martin	D.T.I.
Braithwaite, Dr Martin	ICI Research & Technology
Brown, Mr Alexander	Defence Research Agency
Buckle, Mr Peter	Defence Research Agency
Cameron, Dr Ian	ICI Explosives
Campbell, Miss Mairi	University of Strathclyde
Campbell, Mr Cairns	Asset Trust
Cartwright, Mr Peter	ICI Explosives
Cashman, Mr William	Irish Industrial Explosives Ireland
Chapman, Mr Barry	Halliburton M & S Ltd
Christie, Mr Ian	Ritchies
Clarke, Mr Philip	Bridgwater
Clason, Mr Anders	Swedish Embassy
Clements, Mr Alan	ICI Explosives Europe
Connor, Dr John	Ministry of Defence
Conway, Miss Shiela	Kilwinning
Coyle, Mr David	Ministry of Defence
Crabbe, Mr John	Health & Safety Executive
Crawford, Mr Robert	ICI Explosives
Cunningham, Mr Jim	ICI Explosives
Dalgleish, Dr William	ICI Explosives
De Cooman, Mrs Chantal	FEEM c/o CEFIC Belgium
Dick, Miss Ruby	Saltcoats
Dolan, Mr John	The Royal Society of Chemistry
Dolan, Mrs Mary	Ardrossan
Dunn, Mr Stuart	Ritchies
Elliot, Dr Mark	Defence Research Agency
Evans, Mr Christopher	Defence Research Agency
Farnfield, Mr Robert	University of Leeds
Fitzpatrick, Mr Peter	ICI Explosives
Francis, Mr Peter	Rock Fall Company Ltd
Gallagher, Mr Leslie	Glasgow Airport Ltd
Gaulter, Miss Sally	Defence Research Agency
Gautier, Mr Jean-Jacques	SNPE France
George, Ms Kate	RCAHMS

Glover, Mr Mike   Explosives Industry Group
Gotham, Dr Ray   Exchem plc
Gray, Mr Allan   ICI Explosives Europe
Groves, Mr Mike   Ministry of Defence
Guppy, Ms Nancy   The Royal Society of Chemistry

Hackett, Mr J   ICI Explosives Europe
Hampel, Sir Ronald   ICI
Hancock, Dr David   WS Atkins Mining
Hansson, Dr Anders   Reaction Engines Ltd
Harley, Bailie William   City of Glasgow
Haslett, Mr Ashley   Ulster Industrial Explosives
Hatt, Mr Mark   W J Hatt Limited
Heino, Mr Pekka   Vihtavuori Oy Finland
Heyes, Mr Duncan   Manchester Airport
Higgins, Mr John   Irish Industrial Explosives Ireland
Hodson, Dr John   DERA
Horne, Mr David   Taylor Woodrow Mining Division
Howley, Mr Mike   Exchem Explosives

Jackson, Mr Alex   ECC International Ltd
Jeacocke, Dr James   Exchem plc
Johannes, Mr Gudmar   Nitro Nobel Sweden
Jones, Mr Harry   Defence Research Agency
Joyner, Mr Brian   The Royal Society of Chemistry
Jurvelin, Mr Antii   Vihtavuori Oy Finland

Kelly, Miss Anette   ICI Explosives
Killian, Mr Tony   Irish Industrial Explosives Ireland
Knowles, Mr David   Vibrock Limited
Krebs, Dr Holger   BAM Germany

Lagerkvist, Dr Per   Bofors Explosives AB Sweden
Langer, Mr Stanley   The Royal Society of Chemistry
Larsson, Mr Bernt   Kimit AB Sweden

Lavery, Mr Philip   Asset Trust
Leeming, Dr William   ICI Explosives
Leiper, Mr Graeme   ICI Explosives
Lewis, Dr Robert   Forensic Science Agency
Liddell, Mr David   Bishopton
Lindstrom, Mr Ronny   Bofors Explosives AB Sweden

Mackay, Mr James   Chemfreight Training Ltd
John MacKenzie   Earl of Cromartie
Magnusson, Mr I   Nitro Nobel Sweden
Mather, Mr W   ICI Explosives Europe
McGoff, Mr Peter   Rocklift Co Ltd

McKay, Mr Ian	Health & Safety Executive
Meline, Mr Pierre	Nobel Explosifs France
Mercer, Mr Rodger	RCAHMS
Monson, Dr B	Health & Safety Inspectorate
Moody, Mr Martin	Defence Research Agency
Morton, Mr Neil	Health & Safety Executive
Mullan, Mr Terry	Irish Industrial Explosives Ireland
Murray, Mr Frank	FEEM
Murray, Dr Stephen	RMCS Cranfield University
Murray, Mr Eamond	Albion Drilling Services Ltd
Naylor, Mr John	The Institution of Mining Engineers
Newton, Mr Keith	DERA
Nylund, Mr Torsten	Kimit AB Sweden
Oglethorpe, Dr Miles	RCAHMS
Parker, Mr Vernon	ICI Explosives
Patz, Mr Vivian	Expert Explosives South Africa
Paul, Mr Norman	Defence Research Agency
Pearson, Dr Jack	ICI Explosives
Pettitt, Mr Andrew	Schlumberger Wireline & Testing
Philbin, Mr Simon	Defence Research Agency
Phillips, Mr John	Health & Safety Executive
Pitcher, Mr Bruce	ICI Explosives Europe
Pittam, Dr David	ICI Explosives
Profit, Mr Alexander	ICI Explosives
Reid, Dr. Robin	St. Andrew's Acedemy
Richards, Mr Mark	ECC International Ltd
Ricketts, Brig. Anthony	Asset Trust
Ronay, Mr Chris	Inst of Makers of Explosives USA
Ruxton, Mr Alan	Blue Circle Cement
Salomonsson , Mr Bengt	Norsk Hydro Sweden
Sexstone, Col Peter	Ministry of Defence
Sharp, Dr Michael	AWE Aldermaston
Shove, Mr Graeme	CAMAS Aggregates
Sime, Mr Martin	DERA
Srihari, Mr Hari	ICI Explosives India
Start, Mr David	Exchem Explosives
Steel, Mr Gordon	Rock Fall Company Ltd
Stell, Mr Geoffey	RCAHMS
Stevenston, Mr John	RCAHMS
Strecker, Mr Karl	Irish Industrial Explosives Ireland
Svoboda, Dr Bohumil	Geodyne Ltd. Czech Republic

Temple, Mr Gordon	Exchem Explosives
Thibaudeau, Mr Alain	Exchem plc
Thomson, Dr Bruce	Health & Safety Laboratory
Trautmann, Dr Wilhelm	Dynamit Nobel GmbH Germany
Veljanov, Dr Slnchomil	CESP/Fac of Crim. Sci. Macedonia
von Bohlen und Halbach, Mr Eckbert	
	Bohlen Industrie GmbH Germany
Weir, Mr Thomas	Irish Cement Ltd Ireland
Wellingham, Mrs Elaine	Conference Secretariat
Wharton, Dr Roland	Health & Safety Laboratory
White, Mr Terry	Exchem Explosives
White, Dr Peter	University of Strathclyde
Williamson, Mr Eddie	Health & Safety Executive
Wilson, Mr. Gerald	Scottish Office
Wilson, Dr Rod	Graseby Dynamics

Subject Index

Schering - Plough (Avondale) Co.
Rathdrum, Co. Wicklow Ireland.